普通高等教育"十三五"规划教材

化工专业英语

第二版

丁丽　王志萍　编

化学工业出版社

·北京·

《化工专业英语》(第二版)是为化工及相关专业编写的专业英语教材,旨在提高学生阅读英语科技文献水平。主要内容包括科技英语的特点以及化学反应、化学工程、单元操作、最新技术等共26个单元。每个单元由课文、词汇表、难点注释、课后练习和阅读材料组成。文章内容丰富,语言难度适中,编排深入浅出,循序渐进。书后还附有总词汇表、化学化工常用构词、常用有机基团名称,便于读者查阅和自学。

《化工专业英语》(第二版)可作为化工及相关专业本科、高职高专学生和在职硕士生的专业英语教材,也可作为从事化工生产和产品经销的人员学习英语的参考书。

图书在版编目(CIP)数据

化工专业英语/丁丽,王志萍编. —2版. —北京:化学工业出版社,2019.12 (2024.2重印)
普通高等教育"十三五"规划教材
ISBN 978-7-122-35420-4

Ⅰ.①化… Ⅱ.①丁… ②王… Ⅲ.①化学工业-英语-高等学校-教材 Ⅳ.①TQ

中国版本图书馆CIP数据核字(2019)第231333号

责任编辑:刘俊之　　　　　　　　　　装帧设计:韩　飞
责任校对:张雨彤

出版发行:化学工业出版社(北京市东城区青年湖南街13号　邮政编码100011)
印　　刷:三河市航远印刷有限公司
装　　订:三河市宇新装订厂
787mm×1092mm　1/16　印张12　字数322千字　2024年2月北京第2版第4次印刷

购书咨询:010-64518888　　　　　　　　售后服务:010-64518899
网　　址:http://www.cip.com.cn
凡购买本书,如有缺损质量问题,本社销售中心负责调换。

定　　价:38.00元　　　　　　　　　　　　　　　　版权所有　违者必究

前　言

随着社会对化工专业技术人才素质要求的提高，具备化工专业知识技能并掌握化工专业英语的技术人才越来越受到社会各界的欢迎。《化工专业英语》（Specialized English for Chemical Engineering）是化工专业学生一门非常重要的课程，着重培养学生对英文资料的阅读及写作的能力。为适应新时期对化工专业学生的能力培养目标和综合素质的要求，我们针对化工专业的学生，分析并比较众多的同类教材，博采众长，结合青岛科技大学化工专业的教学实践，编写了本书。

《化工专业英语》（第二版）保留了第一版的特色，修改了上一版中的错漏，在选编内容时紧扣化工专业知识，从适用及专业方面考虑，突出实用、够用和好用的特点。首先介绍了科技英语的特点，课文内容涵盖了化学工程的基础内容和前沿领域，主要内容包括化学反应、化学工程、单元操作及最新技术等共 26 个单元。每个单元由课文、词汇表、难点注释、课后练习和阅读材料组成。选材范围广、词汇全面，所选内容适应性强，覆盖面宽，图文并茂，内容丰富，适宜化工类及相关专业本科、高职高专学生及在职研究生作为教材选用或自学。本书还配有课件（www.cipedu.com.cn）。

由于编者水平有限，书中不妥之处在所难免，敬请使用本书的读者指正，不胜感激。

编　者
2019 年 9 月

目 录

Summarization　Features of English for Science and Technology 1

Lesson 1　Chemical Engineering 7
　　　　Reading Material　What is Chemical Engineering 10

Lesson 2　Chemical Equilibrium and Kinetics 12
　　　　Reading Material　Chemical Kinetics 15

Lesson 3　The Second Law of Thermodynamics 18
　　　　Reading Material　Chemical and Process Thermodynamics 20

Lesson 4　Chemical Reaction Engineering 24
　　　　Reading Material　Reactor Technology 27

Lesson 5　Chlor-Alkali and Related Processes 32
　　　　Reading Material　Sulphuric Acid 36

Lesson 6　Ammonia 40
　　　　Reading Material　Nitric Acid and Urea 43

Lesson 7　Momentum, Heat, and Mass Transfer 46
　　　　Reading Material　Fluid-Flow Phenomena 48

Lesson 8　Heat Transfer 52
　　　　Reading Material　Classification of Heat Transfer Equipment 55

Lesson 9　Distillation 57
　　　　Reading Material　Azeotropic and Extractive Distillation 59

Lesson 10　Gas Absorption 61
　　　　Reading Material　Principle Types of Absorption Equipment 63

Lesson 11　Liquid Extraction 66
　　　　Reading Material　The Industrial Application of Liquid-Liquid Extraction 69

Lesson 12　Surface Chemistry and Adsorption 71

　　　　　　　　Reading Material　Adsorption Isotherms ···································· 73

Lesson 13　Filtration ·· 75
　　　　　　　　Reading Material　Types of Filtration Equipment ·························· 77

Lesson 14　Evaporation, Crystallization and Drying ·· 80
　　　　　　　　Reading Material　Evaporator, Crystallizer, and Dryer ···················· 82

Lesson 15　Computer-Assisted Design of New Process ·· 86
　　　　　　　　Reading Material　Process Design in the 21st Century ···················· 89

Lesson 16　Catalysis ··· 90
　　　　　　　　Reading Material　Classification of Catalysts ································ 92

Lesson 17　Colloid ··· 95
　　　　　　　　Reading Material　Coagulation and Flocculation ·························· 100

Lesson 18　Polymers and Polymerization Techniques ·· 103
　　　　　　　　Reading Material　Petroleum Processing ······································ 108

Lesson 19　Chemical Industry and Environment ·· 113
　　　　　　　　Reading Material　Chemical Process Safety ································ 118

Lesson 20　Vapor-Phase Chromatography ·· 120
　　　　　　　　Reading Material　High Performance Liquid Chromatography and
　　　　　　　　Capillary Electrophoresis ·· 123

Lesson 21　Membranes for Separation Process ·· 130
　　　　　　　　Reading Material　Membranes in Chemical Processing ················ 132

Lesson 22　New Technologies in Unit Operations ·· 135
　　　　　　　　Reading Material　What Is Materials Science and Engineering? ···· 137

Lesson 23　Structure and Nomenclature of Hydrocarbons ···································· 141
　　　　　　　　Reading Material　Nomenclature of Chemical Compounds ············ 148

Lesson 24　Green Science and Technology ·· 151
　　　　　　　　Reading Material　Green Chemistry ·· 154

Lesson 25　Processing and Utilization of Coal ·· 158
　　　　　　　　Reading Material　Summary of Coal Chemical Technology ·········· 160

Lesson 26　Reading and Searching a Patent …………………………………………162
　　　　　　Reading Material　Design Information and Data……………………166

Appendix …………………………………………………………………………170
　　Appendix 1　化学化工常用构词……………………………………………170
　　Appendix 2　常用有机基团…………………………………………………171
　　Appendix 3　总词汇表………………………………………………………172

Summarization

Features of English for Science and Technology

科技文体崇尚严谨周密，概念准确，逻辑性强，行文简练，重点突出，句式严整，少有变化。科技文章文体的特点是：清晰、准确、精练、严密。那么，科技文章的语言结构特色在翻译过程中如何处理，这是进行英汉科技翻译时需要探讨的问题。

一、科技英语的特点

1. 大量使用名词化结构

大量使用名词化结构（Nominalization）是科技英语的特点之一。因科技文体要求行文简洁、表达客观、内容确切、信息量大、强调存在的事实，而非某一行为。

Archimedes first discovered the principle of displacement of water by solid bodies. 阿基米德最先发现固体排水的原理。句中 the principle of displacement of water by solid bodies 系名词化结构，一方面简化了同位语从句，另一方强调 displacement 这一事实。

The rotation of the earth on its own axis causes the change from day to night. 地球绕轴自转，引起昼夜的变化。名词化结构 the rotation of the earth on its own axis 使复合句简化成简单句，而且使表达的概念更加确切严密。

Television is the transmission and reception of images of moving objects by radio waves. 电视通过无线电波发射和接受活动物体的图像。名词化结构 the transmission and reception of images of moving objects by radio waves 强调客观事实，而谓语动词则着重其发射和接受的能力。

2. 广泛使用被动语句

根据英国利兹大学 John Swales 的统计，科技英语中的谓语至少三分之一是被动态。这是因为科技文章侧重叙事推理，强调客观准确。第一、二人称使用过多，会造成主观臆断的印象。因此尽量使用第三人称叙述，采用被动语态，例如：

Attention must be paid to the working temperature of the machine. 应当注意机器的工作温度。而很少说：You must pay attention to the working temperature of the machine. 你们必须注意机器的工作温度。

此外，科技文章将主要信息前置，放在主语部分。这也是广泛使用被动态的主要原因。试观察并比较下列两段短文的主语。

Electrical energy can be stored in two metal plates separated by an insulating medium. Such a device is called a capacitor, and **its** ability to store electrical energy is called capacitance. **It** is measured in farads. 电能可储存在由一绝缘介质隔开的两块金属极板内。这样的装置称为电容器，其储存电能的能力称为电容。电容的单位是法拉。

这一段短文中各句的主语分别为：Electrical energy, Such a device, Its ability to store electrical energy, It (Capacitance)。它们都处于句首的位置，非常醒目。足见被动结构可收简洁客观之效。试比较：

We can store electrical energy in two metal plates separated by an insulating medium. We call such a device a capacitor, or a condenser, and call its ability to store electrical energy capacitance. We measure it in farads.

3. 经常运用非限定动词

如前所述，科技文章要求行文简练，结构紧凑，为此，往往使用分词短语代替定语从句或状语从句；使用分词独立结构代替状语从句或并列分句；使用不定式短语代替各种从句；介词+动名词短语代替定语从句或状语从句。这样可缩短句子，又比较醒目。试比较下列各组句子。

A direct current is a current flowing always in the same direction. 直流电是一种总是沿同一方向流动的电流。

Radiating from the earth, heat causes air currents to rise. 热量由地球辐射出来时，使得气流上升。

A body can move uniformly and in a straight line, there being no cause to change that motion. 如果没有改变物体运动的原因，那么物体将做匀速直线运动。

Vibrating objects produce sound waves, each vibration producing one sound wave. 振动着的物体产生声波，每一次振动产生一个声波。

In communications, the problem of electronics is how to convey information from one place to another. 在通信系统中，电子学要解决的问题是如何把信息从一个地方传递到另一个地方。

Materials to be used for structural purposes are chosen so as to behave elastically in the environmental conditions. 结构材料的选择应使其在外界条件中保持其弹性。

There are different ways of changing energy from one form into another. 将能量从一种形式转变成另一种形式有各种不同的方法。

In making the radio waves correspond to each sound in turn, messages are carried from a broadcasting station to a receiving set. 当无线电波依次对每一个声音作出相应变化时，信息就由广播电台传递到接收机。

4. 频繁使用后置定语

大量使用后置定语也是科技文章的特点之一。常见的结构有以下五种：

（1）介词短语。The forces due to friction are called frictional forces. 由于摩擦而产生的力称之为摩擦力。The data in table 1 show⋯.

（2）形容词及形容词短语。In this factory the only fuel available is coal. 该厂唯一可用的燃料是煤。In radiation, thermal energy is transformed into radiant energy, similar in nature to light. 热能在辐射时，转换成性质与光相似的辐射能。

（3）副词。The temperature of air outside is⋯. 外面的空气的温度是⋯。The force upward equals the force downward so that the balloon stays at the level. 向上的力与向下的力相等，所以气球就保持在这一高度。

（4）单个分词。The results obtained must be checked. 获得的结果必须加以校核。The heat produced is equal to the electrical energy wasted. 产生的热量等于浪费了的电能。

（5）定语从句。During construction, problems often arise which require design changes. 在施工过程中，常会出现需要改变设计的问题。The molecules exert forces upon each other, which depend upon the distance between them. 分子相互间都存在着力的作用，该力的大小取决于它们之间的距离。Very wonderful changes in matter take place before our eyes every day to which we pay little attention. （定语从句 to which we pay little attention 修饰的是 changes，这是一种分隔定语从句。）我们几乎没有注意的很奇异的物质变化每天都在眼前发生。To make an atomic bomb we have to use uranium 235, in which all the atoms are available for fission. 制造原子弹，我们必须用铀235，因为铀235的所有原子都会裂变。

5. 常用特定的句型

科技文章中经常使用若干特定的句型，从而形成科技文体区别于其他文体的标志。例如：**It…that…**结构句型；被动态结构句型；比较结构句型；分词短语结构句型；省略句结构句型等。举例如下：

It is evident that a well lubricated bearing turns more easily than a dry one. 显然，润滑好的轴承，比不润滑的轴承容易转动。

It seems that these two branches of science are mutually dependent and interacting. 看来这两个科学分支是相互依存，相互作用的。

It has been proved that induced voltage causes a **current** to flow in opposition to the force producing **it**. 已经证明，感应电压使电流的方向与产生电流的磁场力方向相反。

It was not until the 19th century that heat was considered as a form of energy. 直到 19 世纪人们才认识到热是能量的一种形式。 Computers may be classified as analog and digital. 计算机可分为模拟计算机和数字计算机两种。

6. 长句使用较多

为了表述一个复杂概念，使之逻辑严密，结构紧凑，科技文章中往往出现许多长句。例如：Major disadvantages of fermentation **compared with petrochemical processes**(过去分词短语作状语)are, firstly, the **time scale**(时间规模), **which usually**(省略了 is) of the **order**(数量级)of days **compared to literally**(简直，几乎)seconds for some catalytic petrochemical reactions, and secondly, the **fact that** the product is usually obtained as a **dilute aqueous solution** (<10% concentration)(稀的水溶液)。

7. 有大量的复合词与缩略词

大量使用复合词与缩略词是科技文章的特点之一，复合词从过去的双词组合发展到多词组合；缩略词趋向于任意构词，这给翻译工作带来一定的困难。例如：full-enclosed 全封闭的（双词合成形容词）；feed-back 反馈（双词合成名词）；work-harden 加工硬化（双词合成动词）；on-and-off-the-road 路面越野两用的（多词合成形容词）；anti-armoured-fighting-vehicle-missile 反装甲车导弹（多词合成名词）；radiophotography 无线电传真（无连字符复合词）；colorimeter 色度计（无连字符复合词）；maths (mathematics)数学（裁减式缩略词）； lab (laboratory)实验室；ft (foot/feet)英尺；cpd (compound)化合物；FM (frequency modulation)调频（用首字母组成的缩略词）；P.S.I. (pounds per square inch) 磅/英寸；SCR (silicon controlled rectifier)可控硅整流器；PVC、PP、PS、MTBE 等等。

二、科技英语汉译时需注意的问题

1. 科技术语的汉译

术语是表示某一专门概念的词语，科技术语就是在科技方面表示某一专门概念的词语。因此翻译时要十分注意，不能疏忽。英语科技术语的特点是词义繁多，专业性强，翻译时必须根据专业内容谨慎处理，稍不注意就会造成很大的错误。如有人把"the newly developed picture tub"（最新研制成功的显像管）错译为"新近被发展的画面管"。如 couple，热电偶，夫妇；column，塔、柱、栏；有多个意思，究竟应译为哪个意思，要从上下文的具体语境去分析判断。

2. 科技英语中倍数增减（包括比较）的汉译

科技英语中倍数增减句型究竟应当如何汉译，在我国翻译界中一直存在着争论，国内出版的一些语法书和工具书所持看法也不尽一致，这就影响了对这种句型的正确翻译。这个问题比较重要，数据上的一倍之差往往会造成不可估量的损失。同时，倍数增减这个问题，在科技英语中又是经常会遇到的。因此，其译法很有必要加以统一。

（1）倍数增加的译法

英语中说"增加了多少倍"，都是连基数也包括在内的，是表示增加后的结果；而在汉语里所谓"增加了多少倍"，则只表示纯粹增加的数量。所以英语里凡表示倍数增加的句型，汉译时都可译成"是……的几倍"，或"比……增加（$n-1$）倍"。现将英语中表示倍数增加的一些表达法及其译法举例如下：

① The production of various stereo recorders has been increased four times as against 1977. 各种立体声录音机的产量比 1977 年增加了三倍。

② The output of colour television receivers increased by a factor of 3 last year. 去年彩色电视接收机的产量增加了二倍。

（2）倍数比较的译法

① "n times + larger than + 被比较对象"，表示其大小"为……的 n 倍"，或"的比……大 $n-1$ 倍"。例如：This thermal power plant is four times larger than that one. 这个热电站比那个热电站大三倍。这是因为英语在倍数比较的表达上，其传统习惯是 larger than 等于 as large as，因此汉译时不能只从字面上理解，将其译为"比……大 n 倍"，而应将其译为"是……的 n 倍"，或"比……大 $n-1$ 倍"。

② "n times + as + 原级 + as + 被比较对象"，表示"是……的 n 倍"。例如：Iron is almost three times as heavy as aluminium. 铁的重量几乎是铝的三倍。

（3）倍数减少的译法

英语中一切表示倍数减少的句型，汉译时都要把它换成分数，而不能按照字面意义将其译成减少了多少倍。因为汉语是不用这种表达方式的，所以应当把它译成减少了几分之几，或减少到几分之几。我们所说的增减多少，指的都是差额，差额应当是以原来的数量为标准，而不能以减少后的数量作标准。英语表示倍数减少时第一种表达方式为："……+ 减少意义的谓语 + by a factor of n 或 by n times"。这种表达法的意思是"成 n 倍地减少"，即减少前的数量比减少后的数量。

The automatic assembly line can shorten the assembling period (by) ten times. 自动装配线能够缩短装配期十分之九。 This metal is three times as light as that one. 这种金属比那种金属轻三分之二。第二种表达方式为"n times + 减少意义的比较级"。

3. 科技英语中部分否定句的汉译

在英语的否定结构中，由于习惯用法问题，其中部分否定句所表示的意思是不能按字面顺序译成汉语的，因此，翻译时要特别注意。英语中含有全体意义的代词和副词如 all，every，both，always，altogether，entirety 等统称为总括词。它们用于否定结构时不是表示全部否定，而只表示其中的一部分被否定。因此，汉译时不能译作"一切……都不"，而应译为"并非一切……都是的"，或"一切……不都是"。例如：

（1）All of the heat supplied to the engine is not converted into useful work. 并非供给热机的所有热量都被转变为有用的功。错译：所有供给热机的热量都没有被转变为有用的功。

（2）Every one cannot do these tests. 并非人人都能做这些试验。错译：每个人都不能做这些试验。

（3）Both instruments are not precise. 两台仪器并不都是精密的。错译：两台仪器都不是精密的。

（4）This plant does not always make such machine tools. 这个工厂并不总是制造这样的机床。错译：这个工厂总是不制造这样的机床。

但是当（总括词 + 肯定式谓语 + 含否定意义的单词……）时，则是表示全部否。例如：All germs are invisible to the naked eye. 一切细菌都是肉眼看不见的。Every design made by her is impossible of execution. 她所做的一切设计都是不能执行的。Both data are incomplete. 两个

数据都不完整。In practice, error sometimes always seems unavoidable. 在实践中，差错有时似乎总是不可避免的。

4. 定语从句的汉译

从汉译的角度来看，英语的定语从句确实要比其他各种从句难些，而且它应用极广，出现的频率很高，科技英语里的许多长句又都离不开定语从句，所以，如何译好定语从句是科技英语汉译工作中的一个重要课题。

语言现象千变万化，定语从句更是如此，因此要想找出一条汉译英语定语从句的规律，确非易事。以下看法，仅供参考。

（1）不论是限制性还是非限制性定语从句，只要是比较短的，或者虽然较长，但汉译后放在被修饰语之前仍然很通顺，一般就放在被修饰语之前，这种译法叫作**逆序合译法**。例如：

The speed of wave is the distance **it** advances per unit time. 波速是波在单位时间内前进的距离。

The light wave that has bounced off the reflecting surface is called the reflected ray. 从反射表面跳回的光波称为反射线。

Stainless steel which is very popular for its resistance to rusting contains large percentage of chromium. 具有突出防锈性能的不锈钢含铬的百分比很高。

（2）定语从句较长，或者虽然不长，但汉译时放在被修饰语之前实在不通顺的就后置，作为词组或分句。这种译法叫做**顺序分译法**。例如：

Each kind of atom seems to have a definite number of "hand" that it can use to hold on to others. 每一种原子似乎都有一定数目的"手"，用来抓牢其他原子 (顺序分译法)。 每一种原子似乎都有一定数目用于抓牢其他原子的手 (逆序合译法)。这句限制性定语从句虽然不长，但用顺序分译法译出的译文要比用逆序合译法更为通顺。

Let AB in the figure above represent an inclined plane the surface of which is smooth and unbending. 设上图中 AB 代表一个倾斜平面，其表面光滑不弯 (顺序分译法) 。设上图中 AB 代表一个其表面为光滑不弯的倾斜平面 (逆序合译法)。

上面两种译法，看来也是用顺序分译法比用逆序合译法更为通顺简明。

（3）定语从句较长，与主句关联又不紧密，汉译时就作为独立句放在主句之后。这种译法仍然是**顺序分译法**。例如：

① Such a slow compression carries the gas through a series of states each of **which** is very nearly an equilibrium state and **it** is called a quasi-static or a "nearly static" process.

这样的缓慢压缩能使这种气体经历一系列的状态，各状态都很接近于平衡状态，所以叫作准静态过程，或"近似静态"过程。

② Friction wears away metal in the moving parts, which shortens **their** working life.

运动部件间的摩擦力使金属磨损，这就缩短了运动部件的使用寿命。

（4）There + be 句型中的限制性定语从句汉译时往往可以把主句中的主语和定语从句溶合一起，译成一个独立的句子。这种译法叫作**溶合法**，也叫拆译法。例如：

① There are bacteria that help plants grow, others that get rid of dead animals and plants by making **them** decay and some that live in soil and make **it** better for growing crops.

有些细菌能帮助植物生长，另一些细菌则通过腐蚀来消除死去的动物和植物，还有一些细菌则生活在土壤里，使土壤变得对种植庄稼更有好处。

② There is a one-seat car which you could learn to drive in fifty minutes.

有一种单座式汽车，五十分钟就能学会驾驶。

5. 科技翻译应符合逻辑

科技翻译不仅仅是个语言问题（词汇、语法、修辞等），它牵涉到许多非语言方面的因素。

逻辑便是其中最活跃、最重要的因素。苏联语言学家巴尔胡达罗夫曾举过这样一个例子：John is in the pen，任何人也不会把句中的 pen 译为笔，而只能译为"牲口圈"，因为"人在钢笔里"是不合事理的。这说明在翻译中常常会碰到需要运用逻辑来判断和解决一些似乎不合逻辑的语言现象。

  Before these metals in their natural state can be converted into useful forms to be of service to man, they must be separated from the other elements or substances with which they are combined. Chemists, who are well acquainted with the properties of metals, have been able to develop processes for separating metals from substances with which they are combined in nature.

  原译：处于天然状态的金属在转换成为人类服务的有用形式之前，必须从和它结合在一起的其他元素或物质中分离出来。化学家们十分熟悉金属的性能。他们已研究出一些方法，把金属从和它在自然界中结合在一起的物质中分离出来．译文将定语从句"who…metals"拆译成独立的句子，译成"化学家们十分熟悉金属的性能"不合逻辑，因为并不是任何化学家都十分熟悉金属的性能。应译为"熟悉金属性能的化学家们已经研究出一些方法，能够……"

三、科技英语的翻译技巧

  要提高翻译质量，使译文达到"准确""通顺""简练"这三个标准，就必须运用翻译技巧。翻译技巧就是在翻译过程中用词造句的处理方法，如词义的引申、增减和词类转换等。

  1. 引申译法

  当英语句子中的某个词按词典的释义直译不符合汉语修辞习惯或语言规范时，则可以在不脱离该英语词本义的前提下，灵活选择恰当的汉语词语或词组译出。例如：We will fix this problem during the recent shut down of the equipment. 我们会在最近的设备停产时解决这一问题。"fix"字典意思为"固定、修理"，这里引申译为"解决、处理"。

  2. 增减词译法

  增词就是在译句中增加或补充英语句子中原来没有或省略了的词语，以便更完善、更清楚地表达英语句子所阐述的内容。在英语句子中，有的词从语法结构上讲是必不可少的，但并无什么实际意义，只是在句子中起着单纯的语法作用；有的词虽有实际意义，但按照字面译出又显多余。这样的词在翻译时往往可以省略不译。

  3. 词类转换

  英语翻译中，常常需要将英语句子中属于某种词类的词，译成另一种词类的汉语词，以适应汉语的表达习惯或达到某种修辞目的。这种翻译处理方法就是转换词性法，简称词类转换。例如：In any case, the performance test have priority. 不管怎样进行，性能测试都要优先。这里将名词"priority"转译为动词"优先"。

  4. 词序处理法

  英汉两种语言的词序规则基本相同，但也存在着某些差别。不同的英语句子，在翻译中的词序处理方式也常常不同。例如：An insufficient power supply makes the motor immovable. 电力不足就会使马达停转。这里将"insufficient power"（不足电力）改序翻译为"电力不足"较为合理。

Lesson 1
Chemical Engineering

Chemical engineering is the development of processes and the design and operation of plants in which materials undergo changes in physical or chemical state on a technical scale. Applied throughout the process industries, it is founded on the principles of chemistry, physics, and mathematics. The laws of physical chemistry and physics govern the practicability and efficiency of chemical engineering operations. Energy changes, deriving from thermodynamic considerations, are particularly important. Mathematics is a basic tool in optimization and modeling. Optimization means arranging materials, facilities, and energy to yield as productive and economical an operation as possible. Modeling is the construction of theoretical mathematical prototypes of complex process systems, commonly with the aid of computers.

Chemical engineering is as old as the process industries. Its heritage dates from the fermentation and evaporation processes operated by early civilizations. Modern chemical engineering emerged with the development of large-scale, chemical-manufacturing operations in the second half of the 19th century. Throughout its development as an independent discipline, chemical engineering has been directed toward solving problems of designing and operating large plants for continuous production.

Manufacture of chemicals in the mid-19th century consisted of modest craft operations. Increase in demand, public concern at the emission of noxious effluents, and competition between rival processes provided the incentives for greater efficiency. This led to the emergence of combines with resources for larger operations and caused the transition from a craft to a science-based industry. The result was a demand for chemists with knowledge of manufacturing processes, known as industrial chemists or chemical technologists. The term chemical engineer was in general use by about 1900. Despite its emergence in traditional chemicals manufacturing, it was through its role in the development of the petroleum industry that chemical engineering became firmly established as a unique discipline. The demand for plants capable of operating physical separation processes continuously at high levels of efficiency was a challenge that could not be met by the traditional chemist or mechanical engineer.

A landmark in the development of chemical engineering was the publication in 1901 of the first textbook on the subject, by George E. Davis, a British chemical consultant. This concentrated on the design of plant items for specific operations. The notion of a processing plant encompassing a number of operations, such as mixing, evaporation, and filtration, and of these operations being essentially similar, whatever the product, led to the concept of unit operations[1]. This was first enunciated by the American chemical engineer Arthur D. Little in 1915 and formed the basis for a classification of chemical engineering that dominated the subject for the next 40 years. The number of unit operations—the building blocks of a chemical plant—is not large. The complexity arises from the variety of conditions under which the unit operations are conducted.

In the same way that a complex plant can be divided into basic unit operations, so chemical reactions involved in the process industries can be classified into certain groups, or unit processes (e.g., polymerizations, esterifications, and nitrations), having common characteristics[2]. This classification into unit processes brought rationalization to the study of process engineering.

The unit approach suffered from the disadvantage inherent in such classifications: a restricted outlook based on existing practice. Since World War II, closer examination of the fundamental phenomena involved in the various unit operations has shown these to depend on the basic laws of mass transfer, heat transfer, and fluid flow. This has given unity to the diverse unit operations and has led to the development of chemical engineering science in its own right; as a result, many applications have been found in fields outside the traditional chemical industry.

Study of the fundamental phenomena upon which chemical engineering is based has necessitated their description in mathematical form and has led to more sophisticated mathematical techniques[3]. The advent of digital computers has allowed laborious design calculations to be performed rapidly, opening the way to accurate optimization of industrial processes. Variations due to different parameters, such as energy source used, plant layout, and environmental factors, can be predicted accurately and quickly so that the best combination can be chosen[4].

Chemical Engineering Functions. Chemical engineers are employed in the design and development of both processes and plant items. In each case, datas and predictions often have to be obtained or confirmed with pilot experiments. Plant operation and control is increasingly the sphere of the chemical engineer rather than the chemist. Chemical engineering provides an ideal background for the economic evaluation of new projects and, in the plant construction sector, for marketing.

Branches of Chemical Engineering. The fundamental principles of chemical engineering underlie the operation of processes extending well beyond the boundaries of the chemical industry, and chemical engineers are employed in a range of operations outside traditional areas. Plastics, polymers, and synthetic fibers involve chemical reaction engineering problems in their manufacture, with fluid flow and heat transfer considerations dominating their fabrication[5]. The dyeing of a fiber is a mass-transfer problem. Pulp and paper manufactures involve considerations of fluid flow and heat transfer. While the scale and materials are different, these again are found in modern continuous production of foodstuffs. The pharmaceuticals industry presents chemical engineering problems, the solutions of which have been essential to the availability of modern drugs. The nuclear industry makes similar demands on the chemical engineer, particularly for fuel manufacture and reprocessing. Chemical engineers are involved in many sectors of the metals processing industry, which extends from steel manufacture to separation of rare metals[6].

Further applications of chemical engineering are found in the fuel industries. In the second half of the 20th century, considerable numbers of chemical engineers have been involved in space exploration, from the design of fuel cells to the manufacture of propellants[7]. Looking to the future, it is probable that chemical engineering will provide the solution to at least two of the world's major problems: supply of adequate fresh water in all regions through desalination of seawater and environmental control through prevention of pollution.

Selected from "English for Chemical Engineers, by Ma Zhengfei etc., Southeast University Press, 2006, 1-4"

New Words

1. thermodynamics [ˈθəməudaiˈnæmiks] n. 热力学
2. prototype [ˈprəutətaip] n. 原型，主型
3. heritage [ˈheritidʒ] n. 遗产，继承物
4. manufacture [mænjuˈfæktʃə] n. 产品，制造
5. emergence [iˈməːdʒəns] n. 出现，浮现
6. craft [kræft] n. 手艺，技艺
7. enunciate [iˈnʌnsieit] v. 明确叙述
8. rationalization [ˌræʃənəlaiˈzeiʃən] n. 合理化
9. foodstuff [ˈfuːdstʌf] n. 食品，粮食
10. desalination [diːˌsæliˈneiʃən] n. 脱盐
11. pollution [pəˈluːʃən] n. 污染

Notes

1. The notion of a processing plant encompassing a number of operations, such as mixing, evaporation, and filtration, and of these operations being essentially similar, whatever the product, led to the concept of unit operations. 参考译文：注意到加工厂包括的一系列操作，如混合、蒸发、过滤，无论产物是什么，这些操作都基本相同，从而导致了单元操作的概念。

2. In the same way that a complex plant can be divided into basic unit operations, so chemical reactions involved in the process industries can be classified into certain groups, or unit processes (e.g., polymerization, esterifications, and nitrations), having common characteristics. 参考译文：同复杂的工厂可划分为基本的单元操作一样，过程工业中涉及的化学反应也可分成一定的单元过程（如聚合、酯化和硝化），它们具有共同的特性。本句中，group 和 unit process 具有相同含义，前者为普通用词，后者为科技用词。在科技文章中，常有此种情况出现，注意此类现象，可帮助理解。

3. Study of the fundamental phenomena upon which chemical engineering is based has necessitated their description in mathematical form and has led to more sophisticated mathematical techniques. 参考译文：研究化工依赖的基本现象需采用数学形式来描述，并借助复杂的数学技术来解决。

4. Variations due to different parameters, such as energy source used, plant layout, and environmental factors, can be predicted accurately and quickly so that the best combination can be chosen. 参考译文：如所用的能量来源、工厂布置和环境因素这样的不同参数引起的变化可正确和快速地得到预测，就可能选择出最佳的组合。

5. Plastics, polymers, and synthetic fibers involve chemical reaction engineering problems in their manufacture, with fluid flow and heat transfer considerations dominating their fabrication. 参考译文：塑料、聚合物和合成纤维在生产中涉及化学反应工程问题，其中流体流动和传热是生产中主要考虑的因素。

6. rare metals 为稀有金属，而 rare earth 为稀土。

7. In the second half of the 20th century, considerable numbers of chemical engineers have been involved in space exploration, from the design of fuel cells to the manufacture of propellants. 参考译文：20世纪下半叶，从燃料电池的设计到推进剂的生产，相当数量的化学工程师参与了空间的探索。

Exercise

1. Put the following into Chinese:

   thermodynamics   manufacture   craft   foodstuff

desalination	mathematics	evaporation	filtration
rare metal	telecommunication	unglamorous	definition

2. Put the following into English:

主型	出现	明确叙述	合理化
技艺	污染	单元操作	合成纤维

3. Comprehension and toward interpretation

   a. what are chemical engineering and its content?
   b. what concept is the landmark in the development of chemical engineering?
   c. what are the basic laws of chemical engineering science?
   d. Name the functions and branches of chemical engineering you know.

Reading Material

What is Chemical Engineering

Society can associate civil engineers with huge new building complexes and bridges, electronic and electrical engineers with telecommunications and power generation, and mechanical engineers with advanced machinery and automobiles. However, chemical engineers have no obvious monuments which create an immediate awareness of the discipline in the public mind. Nevertheless, the range of products in daily use which are efficiently produced as a result of the application of chemical engineering expertise is enormous. The list given in Table 1-1 is not exhaustive, and any reader who grasps the key element, which involves the conversion of raw materials into a useful product, will be able to extend it. Although the products are unglamorous, the creation and operation of cost-effective processes to produce them is often challenging and exciting.

The term "chemical engineer" implies that the person is primarily an engineer whose first professional concern is with manufacturing processes—making something, or making some process work. The adjective "chemical" implies a particular interest in processes which involve chemical changes. While the main term is correct, the adjective is too restrictive and the literal definition will not suffice. Taken at face value, it would exclude many areas in which chemical engineers have made their mark, for example, textiles, nuclear fuels and the food industry. Thus the Institution of Chemical Engineers defines chemical engineering as "that branch of engineering which is concerned with processes in which materials undergo a required change in composition, energy content or physical state: with the means of processing; with the resulting products, and with their application to useful ends". It is perhaps too presumptuous to insist that the term "process engineer" should replace the term "chemical engineer", and so the two will be used synonymously.

It should also be noted that large-scale processes involving biological systems (such as waste water treatment and production of protein) fit the definition as well as traditional chemical processes such as the production of fertilizers and pharmaceuticals.

The work of chemical engineers will be examined by way of four case studies in the second part of this chapter, but to complete the definition, explicit mention of the concern that process operations be both safe and economic must be made.

A jocular, helpful, but very incomplete description is that, "a chemical engineer is a chemist who is aware of money". Although this neglects many, if not most, aspects of a chemical engineer's training, it does illustrate one important facet of any engineer's work. When working on a large scale, the cost of equipment and raw materials are more important than the cost of manpower. While the research chemist might use aqueous potassium hydroxide to neutralize acids, because it is pure and readily available, the chemical engineer will specify a cheaper alternative, provided that it serves the same purpose. Two obvious substitutes are aqueous sodium hydroxide, which is available at less than a tenth of the cost, or calcium hydroxide, which is even cheaper, but harder to handle. In choosing between these cheaper alternatives, an engineer has to balance the cost of handling a slurry (calcium hydroxide is sparingly soluble) against the higher price of sodium hydroxide.

Table 1-1 A selection of everyday products whose manufacture involves the application of chemical engineering

Process product grouping or production
1. Household products in daily use
2. Health care products
3. Automotive fuels/Petroleum refining
4. Other chemicals in daily use
5. Horticultural products
6. Metals
7. Polymerization, extrusion and molding of thermoplastics
8. Polymerization, production and spinning of synthetic fibers
9. Electronics
10. Fats and oils
11. Fermentation
12. Dairy products
13. Gas treatment and transmission
Some of the more familiar examples
1. Detergents, polishes, disinfectants
2. Pharmaceuticals, toiletries, antiseptics, anesthetics
3. Petrol, diesel, lubricants
4. Latex paints, rubber, anti-freeze, refrigerants, insulation materials
5. Fertilizers, fungicides, insecticides
6. Steel manufacture, zinc production
7. Washing-up bowls, baths, insulation for cables, road signs, children's toys
8. Clothes, curtains, sheets, blankets
9. Raw materials, silicon, gallium arsenide, etchants, dopants
10. Salad and cooking oils, margarine, soap
11. Beer, certain antibiotics such as penicillin, yoghurts
12. Milk, butter, cheese, baby food
13. Gas for heating and cooking

Selected from "English for Chemical Engineers, by Ma Zhengfei etc., Southeast University Press, 2006, 5-6"

Lesson 2
Chemical Equilibrium and Kinetics

A major objective of chemist is to understand chemical reactions, to know whether under a given set of conditions two substances will react when mixed, to determine whether a given reaction will be exothermic or endothermic, and to predict the extent to which a given reaction will proceed before equilibrium is established. An equilibrium state, produced as a consequence of two opposing reactions occurring simultaneously, is a state in which there is no net change as long as there is no change in conditions. In this lesson it will be shown how one predict the equilibrium state of chemical systems from thermodynamic data, and conversely how the experimental measurements states provide useful thermodynamic data. Thermodynamics alone cannot explain the rate at which equilibrium is established, nor does it provide details of the mechanism by which equilibrium is established. Such explanations can be developed from considerations of the quantum theory of molecular structure and from statistical mechanics.

To appreciate fully the nature of the chemical equilibrium state, it is necessary first to have some acquaintance with the factors which influence reaction rates. The factors which influence the rates of a chemical reaction are temperature, concentrations of reactants (or partial pressures of gaseous reactants), and presence of a catalyst. In general, for a given reaction the higher the temperature, the faster the reaction will occur. The concentrations of reactants or partial pressure of gaseous reactants will affect the rate of reaction, an increase in concentration or partial pressure increases the rate of most reactions. Substances which accelerate a chemical reaction but which themselves are not used up in the reaction are called catalysts.

Dynamic Equilibrium

In many cases, direct reactions between two substances appear to cease before all of either starting material is exhausted. Moreover, the products of chemical reactions themselves often react to produce the starting materials. For example, nitrogen and hydrogen combine at 500℃ in the presence of a catalyst to produce ammonia:

$$N_2 + 3H_2 = 2NH_3$$

At the same temperature and in the presence of the same catalyst, pure ammonia decomposes into nitrogen and hydrogen:

$$2NH_3 = 3H_2 + N_2$$

For convenience, these two opposing reactions are denoted in one equation by use of a double arrow:

$$N_2 + 3H_2 \rightleftharpoons 2NH_3$$

The reaction proceeding toward the right is called the forward reaction; the other is called the reverse reaction.

If either ammonia or a mixture of nitrogen and hydrogen is subjected to the above conditions, a mixture of all three gases will result. The rate of reaction between the materials which were introduced into the reaction vessel will decrease after the reaction starts; because their concentrations are decreasing, conversely, after the start of the reaction the material being produced will react faster, since there will be more of it. Thus the faster forward reaction becomes slower, and the slower reverse reaction speeds up. Ultimately the time comes when the rates of the forward and reverse reactions become equal, and there will be no further net change[1]. This situation is called equilibrium. Equilibrium is a dynamic state because both reactions are still proceeding; but since the two opposing reactions are proceeding at equal rates, no net change is observed.

All chemical reactions ultimately proceed toward equilibrium. In a practical sense, however, some reactions go so far in one direction that the reverse reaction cannot be detected, and they are said to go to completion. The principles of chemical equilibrium apply even to these, and it will be seen that for many of them, the extent of reaction can be expressed quantitatively[2].

Equilibrium Constants

Equilibrium is a state of dynamic balance between two opposing processes. For a general reaction at a given temperature:

$$A + B \rightleftharpoons C + D$$

At the point of equilibrium, the following ratio must be a constant:

$$K = \frac{[C][D]}{[A][B]}$$

The constant, K, is called the equilibrium constant of the reaction. It has a specific value at a given temperature. If the concentration of any of the components in the system at equilibrium is changed, the concentrations of the other components will change in such a manner that the defined ratio remains equal to K as long as the temperature does not change[3]. The equilibrium constant expression quantitatively defines the equilibrium state.

More generally, for the reversible reaction

$$aA + bB \rightleftharpoons cC + dD$$

the equilibrium constant expression is written as follows:

$$K = \frac{[C]^c[D]^d}{[A]^a[B]^b}$$

By convention, the concentration terms of the reaction products are always placed in the numerator of the equilibrium constant expression. It should be noted that the exponents of the concentration terms in the equilibrium constant expression are the coefficients of the respective species in the balanced chemical equation.

Chemical Kinetics

When a system is in the equilibrium state, the rate of the forward reaction is identical to the rate of the reverse reaction. It is important to know just how fast a reactant is being used up in a process, or the speed with which a product is being formed. It is also important to have detailed information about rates of reactants in order to test theories and mechanisms for various kinds of chemical processes.

Experiments show that a number of reaction variables affect reaction rates:

Temperature. The rates of chemical reactions are temperature-dependent. Therefore, it is common practice when studying the rate in the laboratory to carry out reactions at constant temperature (isothermally), thus eliminating one variable[4].

Pressure and Volume. Pressure is important in a kinetic consideration of gas phase reactions. Usually, volume is fixed by running the reaction in a container of fixed dimensions. For solid and liquid state reactions, pressure is usually atmospheric, and the volume of the reacting system is relatively unimportant because there is little change in volume.

Concentration. At any particular temperature, the rates of most chemical reactions are functions of the concentrations of one or more of the components of the system. In practice, it is usually the concentration of the reactants that are used in determining the overall rates of reaction[5].

Catalyst. Any substance that affects the rate of a chemical reaction but cannot be identified as a product or reactant is said to be a catalyst. Catalyst may accelerate the rate, but we usually refer to decelerating catalysts as inhibitors.

The order of a chemical reaction is given by the number of atomic or molecular species whose concentrations directly determine the reaction rate.

The rate of hydrolysis of acetate in water is directly proportional to the ethyl acetate concentration, the reaction is said to be first order.

Selected form "Specialized Chemical English (unformal published) (volumn two), by Cui Bo etc., 1994, 66-76"

New Words

1. kinetics [kaiˈnetiks] n. 动力学
2. equilibrium state 平衡状态
3. opposing [əˈpəuziŋ] n. 相反，相对，反抗
4. datum [ˈdeitəm] （复 data [ˈdeitə]) n. 资料，论据
5. molecular structure 分子结构
6. quantum [ˈkwɔntəm] n. 量子
7. quantum theory 量子论
8. statistical mechanics [stəˈtistikəl miˈkæniks] 统计力学
9. acquaintance [əˈkweintns] n. 熟悉，认识 (with)
10. partial pressure [ˈpɑːʃəlˈpreʃə] 分压
11. affect [əˈfekt] vt. 影响
12. accelerate [ækˈseləreit] vt. 加速
13. dynamic [daiˈnæmik] a. 动力的
14. cease [ˈsiːs] v. 停止，终止
15. forward reaction 正反应
16. reverse reaction 逆反应
17. subject [ˈsʌbdʒikt] vt. 遭受，蒙受（to）
18. ultimately [ˈʌltimitli] ad. 最后，最终
19. specific value 比值
20. term [təːm] n. (比例或方程的)项
21. numerator [ˈnjuːməreitə] n. (分数中的)分子
22. exponent [eksˈpəunənt] n. 指数
23. coefficient [ˌkəuiˈfiʃənt] n. 系数，率
24. test [test] vt. 检验，验证，试验
25. isothermal [ˌaisəuˈθəːməl] a. 等温线的
26. decelerate [diːˈseləreit] v. 减速，减慢
27. order [ˈɔːdə] n. 级数
28. rate law 速度定律

Phrases

1. as a consequence of 由于……（结果）
2. as long as… 只要
3. not…, …nor 不……，也不……
4. have acquaintance with… 熟悉，认识
5. use up 消耗掉，用完
6. in a practical sense 实际上
7. in such a manner that… 以这样的方式以致……
8. by convention 按照惯例
9. it is common practice (+ inf.) 通常的做法是……
10. over-all 总的，全部的
11. an educated guess 有根据的推测
12. refer to…as… 把……称作

Notes

1. Ultimately the time comes when the rates of the forward and reverse reactions become equal, and there will be no further net change. 系主从复合句，主句是"Ultimately the time comes"，"when

the rates…net change"是由 when 引导的、由 and 连接的两个并列的定语从句，它们修饰主句中的主语 the time。参考译文："最后，正反应和逆反应速率变得相等，且不再有进一步的净变化"。

2. The principles of chemical equilibrium apply even to these, and it will be seen that for many of them, the extent of reaction can be expressed quantitatively. 该句为 and 连接的并列复合句，第一个分句中的 these 指上句中提到的反应，即"some reactions go so…to completion"。参考译文："化学平衡原理甚至也适用于完全反应，且许多这类反应其反应程度可以定量地表达"。

3. If the concentration of any of the components in the system at equilibrium is changed, the concentrations of the other components will change in such a manner that the defined ratio remains equal to K as long as the temperature does not change. 系主从复合句，if 引导的是条件状语从句，主句是"the concentration of the other…句末"，主句中"in such a manner"为方式状语，"that the defined … not change"为修饰 manner 的定语从句，此定语从句中又带有一个条件状语从句，即"as long as … not change"。参考译文："如果平衡时体系中任意组分的浓度改变，那么只要温度不变，其他组分的浓度将以保持相等 K 的固定比例的方式变化"（意译）。

4. Therefore, it is common practice when studying the rate in the laboratory to carry out reactions at constant temperature (isothermally), thus eliminating one variable. 系主从复合句，主句是"it is common practice to carry out …句末"，其中"thus eliminating one variable"是结果状语，一般 thus + 现在分词的短语在句中常表示结果。"when studying the rate in the laboratory"为时间状语。

5. In practice, it is usually the concentration of the reactants that are used in determining the overall rates of reaction. 此句是强调句型，句中强调了主语"the concentration of the reactions"。参考译文："实际上，一般是用反应物的浓度测定反应的总速率"。

Exercise

1. Put the following into Chinese:

kinetics	forward reaction	use up	correlation
opposing	numerator	distinguish	behaviour
accelerate	isothermal	acquaint	vital

2. Put the following into English:

平衡态	分压	系数	雷诺数
量子	动力学	受益	论点
统计力学	比值	阐明	可行

Reading Material

Chemical Kinetics

Without chemical reaction our world would be a barren planet. No life of any sort would exist. Even if the fundamental reactions involved in life processes did exist without other chemical reactions taking place around us, our lives would be extremely different from what they are today. There would be no fire for warmth and cooking, no iron and steel with which to fashion even the crudest implements, no synthetic fibers for clothing, and no engines to power our vehicles.

One feature that distinguishes the chemical engineer from other types of engineers is the ability to analyze systems in which chemical reactions are occurring and to apply the results of his analysis in a manner that benefits society. Consequently, chemical engineers must be well acquainted with the fundamentals of chemical kinetics and the manner in which they are applied in chemical reactor design.

Chemical Kinetics deals with quantitative studies of the rates at which chemical processes occur, the factors on which these rates depend, and the molecular acts involved in reaction processes. A description of a reaction in terms of its constituent molecular acts is known as the mechanism of the reaction. Physical and organic chemists are primarily interested in chemical kinetics for the light that it sheds on molecular properties. From interpretations of macroscopic kinetic data in terms of molecular mechanisms, they can gain insight into the nature of reacting systems, the processes by which chemical bonds are made and broken, and the structure of the resultant product. Although chemical engineers find the concept of a reaction mechanism useful in the correlation, interpolation, and extrapolation of rate data, they are more concerned with application of chemical kinetics in the development of manufacturing processes.

Chemical engineers have traditionally approached kinetics studies with the goal of describing the behavior of reacting systems in terms of macroscopically observable quantities such as temperature, pressure, composition, and Reynolds number. This empirical approach has been very fruitful in that it has permitted chemical reactor technology to develop to a point that far surpasses the development of theoretical work in chemical kinetics.

The dynamic view point of chemical kinetics may be contrasted with the essentially static viewpoint of thermodynamics. A kinetic system is a system in unidirectional movement toward a condition of thermodynamic equilibrium. The chemical composition of the system changes continuously with time. A system that is in thermodynamic equilibrium, on the other hand, undergoes no net change with time. The thermodynamicist is interested only in the initial and final states of the system and is not concerned with the time required for the transition of the molecular processes involved therein; the chemical kineticist is concerned primarily with these issues.

In principle one can treat the thermodynamics of chemical reactions on a kinetic basis by recognizing that the equilibrium condition corresponds to the cases where the rates of the forward and reverse reactions are identical. In this sense kinetics is the more fundamental science. Nonetheless, thermodynamics provided much vital information to the kineticist and to the reactor designer. In particular, the first step in determining the economic feasibility of producing a given material from a given reactant feed stock should be the determination of the product yield at equilibrium at the conditions of the reactor outlet. Since this composition represents the goal toward which the kinetic process is moving, it places a maximum limit on the product yield that may be obtained. Chemical engineers must also use thermodynamics to determine heat transfer requirements for proposed reactor configurations.

Selected form "*Specialized Chemical English (unformal published) (the first volumn), by Cui Bo etc., 1994, 67-70*"

New Words

1. kinetics [kaiˈnetiks] *n.* 动力学
2. barren [ˈbærən] *a.* 不毛的，贫瘠的
3. cook [ˈkuk] *vt.* 烹调
4. implement [ˈimplimənt] *n.* 工具，器具
5. vehicle [ˈviːikl] *n.* 车辆
6. distinguish [disˈtiŋgwiʃ] *vt.* 区别

7. benefit ['benifit]   vt. 有益于
                        vi. 受益
                        n. 利益，好处
8. acquaint [ə'kweint]   vt. 使认识，使熟悉
   acquaint oneself with（或 of） 使自己知道（熟悉）
9. shed [ʃed]   vt. 放射，散发，流出，流下
   shed light on   阐明，把……弄明白
10. correlation [kɔri'leiʃən]   n. 相互关系，伴随关系，关联（作用）
11. interpolation [in,tɜ:pə'leiʃən]   n. 内插法，插入，解释
12. extrapolation [ekstræpə'leiʃən]   n. 外推法，推断，推知

13. behavior [bi'heiviə]   n. 性能，性质，行为
14. Reynolds number   雷诺数
15. empirical [em'pirikəl]   a. 经验（上）的
16. dynamic [dai'næmik]   a. 动力（学）的
17. unidirectional ['ju:nidi'rekʃənl]   a. 单向性的
18. issue ['isju:]   n. 问题，论点，流出物，流出
19. identical [ai'dentikəl]   a. 相同的，完全相同的
20. nonetheless [nʌnðə'les]   ad. 仍然，不过（=nevertheless）
21. vital ['vaitl]   a. 必需的，生命的
22. kineticist [kai'netisist]   n. 动力学家
23. feasibility [fi:zə'biliti]   n. 可行，可实行

Lesson 3
The Second Law of Thermodynamics

Thermodynamics is concerned with transformations of energy, and the laws of thermodynamics describe the bounds within which these transformations are observed to occur. The first law, stating that energy must be conserved in all ordinary processes, has been the underlying principle of the preceding chapters. The first law imposes no restriction on the direction of energy transformations. Yet, all our experience indicates the existence of such a restriction. To complete the foundation for the science of thermodynamics, it is necessary to formulate this second limitation. Its concise statement constitutes the second law.

The differences between the two forms of energy, heat and work, provide some insight into the second law. These differences are not implied by the first law. In an energy balance both work and heat are included as simple additive terms, implying that one unit of heat, such as a joule or BTU, is equivalent to the same unit of work. Although this is true with respect to a energy balance. Experience teaches that there is a difference in quality between heat and work. This experience can be summarized by the following facts.

First, the efficiency of the transformation from one form of work to another such as electrical to mechanical as accomplished in an electric motor, can be made to approach 100 percent as closely as is desired. One needs merely to exert more and more care in eliminating irreversibilities in the apparatus. On the other hand, efforts to convert energy transferred to a system as heat into any of the forms of work show this regardless of improvements in the machines employed. The conversion is limited to low values (40 percent is an approximate maximum). These efficiencies are so low, in comparison with these obtained for the transformation of work from one form to another, that there can be no escape from the conclusion that there is an intrinsic difference between heat and work, in the reverse direction, the conversion of work into heat with 100 percent efficiency is very common. Indeed, efforts are made in nearly every machine to eliminate this conversion, which decrease efficiency of operation. These facts lead to the conclusion that heat is a less versatile or more degraded form of energy than work. Work might be termed energy of a higher quality than heat.

To draw further upon our experience, we know that heat always flows from a higher temperature level to a lower one, and never in the reverse direction. This suggests that heat itself may be assigned a characteristic quality as well as quantity, and that this quality depends upon temperature, the relation of temperature to the quality of heat is evident from the increased in efficiency with which heat may be converted into work as the temperature of the source is raised. For example, the efficiency, or work output per unit of fuel burned, of a stationary power plant increases as the temperature of the steam in the boiler and superheater rises.

Statements of the Second Law

The observations just described are results of the restriction imposed by the second law on the direction of actual processes. Many general statements may be made which described this restriction and hence, serve as statements of the second law. Two of the most common are:

1. No apparatus can operate in such a way that its only effect (in system and surroundings) is to convert heat absorbed by a system completely into work.

2. Any process which consists solely in the transfer of heat from one temperature to a higher one is impossible.

Statement 1 does not imply that heat cannot be converted into work, but does mean that changes, other than those resulting directly from the conversion of heat into work, must occur in either the system or surroundings. Consider the case of an ideal gas in a cylinder and piston assembly expanding reversibly at constant temperature. Work is produced in the surroundings (consider the gas as the system) equal to the integral of the pressure times the change in volume. Since the gas is ideal, $u=0$ then, according to the first law the heat absorbed by the gas from the surrounding is equal to the work produced in the surroundings because of the reversible expansion in the gas. At first this might seem to be a contradiction of statement 1, since in surroundings the only result has been the complete conversion of heat into work. However the second law statement requires that there also be no change in the system, a requirement which has not been met in this example. Since the pressure of the gas decreased, this process cannot be continued indefinitely. The pressure of the gas would soon reach that of the surroundings, and further expansion would be impossible. Therefore, a method of continuously producing work from heat by this method would fail, if the original state of the system were restored in order to comply with the requirements of statements 1, it would be necessary to take energy from the surroundings in the form of work in order to compress the gas back to its original pressure. At the same time energy as heat would be transferred to the surroundings in order to maintain constant temperature. This reverse process would require just the amount of work gained from the expansion; hence the net work produced would be zero. From this discussion it is evident that statement 1 might be expressed in an alternative way, viz: 1a. it is impossible to convert the heat absorbed completely into work in a cyclical process.

The term cyclical requires that the system be restored periodically to its original state. In the previous example the expansion and compression back to the original state constitute a complete cycle. If the process is repeated, it becomes a cyclical process. The restriction to a cyclical process in statement 1a amounts to the same limitation as that introduced by the words only effect in statement 1.

The second law does not prohibit the production of work from heat, but it does place a limitation upon the efficiency of any cyclic process. The partial conversion of heat into work forms the basis for all commercial plants for the production of power (water power is an exception). The development of a quantitative expression for the efficiency of this process is the next objective in the treatment of the second law.

Selected form "Specialized Chemical English (unformal published) (volumn two), by Cui Bo etc., 1994, 82-87"

New Words

1. concise [kən'sais] a. 简明的，扼要的，短的
2. insight 了解
3. BTU 英制热量单位
4. versatile ['və:sətail] a. 多方面的，多才多艺的；万用的，通用的
5. impose [im'pəuz] v. 征税；把……强加给……
6. viz: (videlicet [vi'di:liset] 之略) 即，就是

Exercise

1. Put the following into Chinese:

   insight   be good for   reboiler   validity
   BTU   shaft   centrifugal   heretofore
   versatile   potential   inductive   diminish

2. Put the following into English:

   简明的   宏观的   冷凝器   可靠性
   就是   微观的   自发的   绝对地
   熵   可逆的   回流   权威的

Reading Material

Chemical and Process Thermodynamics

Before committing a great deal of time and effort to the study of a subject, it is reasonable to ask the following two questions: what is it? What is it good for? Regarding thermodynamics, the second question is more easily answered, but an answer to the first is essential to an understanding of the subject. Although it is doubtful that many experts or scholars would agree on a simple and precise definition of thermodynamics, necessity demands that a definition be attempted. However, this is best accomplished after the applications of thermodynamics have been discussed.

1. Applications of Thermodynamics

There are two major applications of thermodynamics, both of which are important to chemical engineers:

(ⅰ) The calculation of heat and work effects associated with processes as well as the calculation of the maximum work obtainable from a process or the minimum work required to drive a process.

(ⅱ) The establishment of relationships among the variables describing systems at equilibrium.

The first application is suggested by the name thermodynamics, which implies heat in motion. Most of these calculations can be made by the direct implementation of the first and second laws. Examples are calculating the work of compressing a gas, performing an energy balance on an entire process or a process unit, determining the minimum work of separating a mixture of ethanol and water, or evaluating the efficiency of an ammonia synthesis plant.

The application of thermodynamics to a particular system results in the definition of useful properties and the establishment of a network of relationships among the properties and other variables such as pressure, temperature, volume, and mol fraction. Actually, application 1 would not be possible unless a means existed for evaluating the necessary thermodynamic property changes required in implementing the first and second laws. The property changes are calculated from experimentally determined data via the established network of relationships. Additionally, the network of relationships among the variables of a system allows the calculation of values of variables which are either unknown or difficult to determine experimentally from variables which are either available or easier to measure. For example, the heat of vaporizing a liquid can be calculated from measurements of the vapor pressure at several temperatures and the densities of the liquid and vapor phases at several temperatures, and the maximum conversion obtainable in a chemical reaction at any temperature can be calculated from calorimetric measurements performed on the individual substances participating in the reaction.

2. The Nature of Thermodynamics

The laws of thermodynamics have an empirical or experimental basis, and in the delineation of its applications the reliance upon experimental measurement stands out. Thus, thermodynamics might be broadly defined as a means of extending our experimentally gained knowledge of a system or as a framework for viewing and correlating the behavior of the system. To understand thermodynamic, it is essential to keep an experimental perspective, for if we do not have a physical appreciation for the system or phenomenon studied, the methods of thermodynamics will have little meaning. We should always ask the following questions: How is this particular variable measured? How, and from what type of data, is a particular property calculated?

Because of its experimental foundation, thermodynamics deals with macroscopic properties, or properties of matter in bulk, as opposed to microscopic properties which are assigned to the atoms and molecules constituting matter. Macroscopic properties are either directly measurable or calculable from directly measurable properties without recourse to a specific theory. Conversely, while microscopic properties are ultimately determined from experimental measurements, their authenticity depends on the validity of the particular theory applied to their calculation. Herein lies the power and authority of thermodynamics: Its results are independent of theories of matter and are thus respected and confidently accepted.

In addition to the certitude accorded its results, thermodynamics enjoys a broad range of applicability. Thus, it forms an integral part of the education of engineers and scientists in many disciplines. Nevertheless, this panoramic scope is often unappreciated because each discipline focuses only on the few applications specific to it. Actually, any system which can exist in observable and reproducible equilibrium states is amenable to the methods of thermodynamics. In addition to fluids, chemically reacting systems, and systems in phase equilibrium, which are of major interest to chemical engineers, thermodynamics has also been successfully applied to systems with surface effects, stressed solids, and substances subjected to gravitational, centrifugal, magnetic, and electric fields.

Through thermodynamics the potentials which define and determine equilibrium are identified and quantified. These potentials also determine the direction in which a system will move as well as the final state it will reach but offer no information concerning the time required to attain the final state. Thus, time is not a thermodynamic variable, and the study of rates is outside the bounds of thermodynamics except in the limit as the system nears equilibrium. Here rate expressions should be thermodynamically consistent.

The experiments and observations on which the laws of thermodynamics are based are neither grand nor sophisticated. Also, the laws themselves are stated in rather pedestrian language. Yet, from this apparently unimpressive beginning a grand structure has evolved which is a tribute to the inductive powers of the human mind. This never fails to inspire awe in the thoughtful and serious student and has led Lewis[1] and Randall to refer to thermodynamics as a cathedral of science. The metaphor is well chosen for in addition to technical accomplishment and structural integrity one also sees beauty and grandeur. It is no small wonder that the study of thermodynamics can be technically rewarding, intellectually stimulating, and, for some, a pleasurable experience.

3. The Thermodynamic Laws

The First Law. The first law of thermodynamics is simply a statement of the conservation of energy. As shown in Fig.3-1, the sum of all the energy leaving a process must equal the sum of all the energy entering, in the steady state. The laws of conservation of mass and energy are followed implicitly by engineers designing and operating processes of all kinds. Unfortunately, taken by itself, the first law has led to much confusion when attempting to evaluate process efficiency. People talk of energy conservation being an important effort, but in fact, no effort is required to conserve energy—it is naturally conserved.

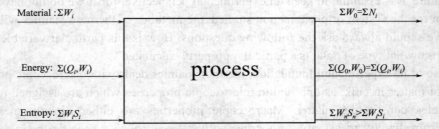

Fig. 3-1   Mass and energy are naturally conserved but entropy is created in every process.

The conclusions which can be drawn from the first law are limited because it does not distinguish among the various energy forms. Shaft work introduced by a reflux pump will leave a column as heat to the condenser just as readily as will heat introduced at the reboiler. Some engineers have fallen into the trap of lumping all forms of energy together in attempting to determine process efficiency. This is obviously not justified—the various energy forms have different costs.

The Second Law. There are many different statements of the second law as applied to cycles in which heat is converted into work. At this point, a more general statement is desirable: The conversion of energy from one form to another always results in an overall loss in quality. Another is: All systems tend to approach equilibrium (disorder). These statements point out the difficulty in expressing the second law. It cannot really be done satisfactorily without defining another term describing quality or disorder.

The first law requires the conservation of energy and gives equal weight to all types of energy changes. While no process is immune to its authority, this law does not recognize the quality of energy nor does it explain why spontaneously occurring processes are never observed to spontaneously reverse themselves. The repeatedly confirmed observation that work may be completely converted into heat but that the reverse transformation never occurs quantitatively leads to the recognition that heat is a lower quality of energy. The second law with its origins deeply rooted in the study of the efficiency of heat engines recognizes the quality of energy. Through this law the existence of a heretofore unrecognized property, the entropy, is revealed, and it is shown that this property determines the direction of spontaneous change. The second law in no way diminishes the authority of the first law; rather it extends and reinforces the jurisdiction of thermodynamics.

The Third Law. The third law of the thermodynamics provides an absolute scale of values for entropy by stating that for changes involving only perfect crystalline solids at absolute zero, the changes of the total entropy is zero. This law enables absolute values to be stated for entropies.

Selected from "Chemical and Process Thermodynamics, B.G.Kyle, Prentice-Hall, Inc., 1984"

Lesson 3 The Second Law of Thermodynamics

New Words

1. (be) good for   对……适用，有效，有利，有好处
2. delineation [dilini'eiʃən] n. 描述，叙述
3. macroscopic [ˌmækrəu'skɔpik] a. 宏观的；肉眼可见的
4. microscopic [ˌmaikrə'skɔpik] a. 微观的，微小的；显微镜的
5. authenticity [ɔ:θen'tisiti] n. 可靠性，真实性
6. validity [və'liditi] n. 有效，合法性；正确，确实
7. certitude ['sə:titju:d] n. 确实，必然性
8. panoramic [pænə'ræmik] a. 全景的，全貌的
9. amenable [ə'mi:nəbl] a. 服从的，适合于……的（to）
10. centrifugal [sen'trifjugəl] a. 离心力的，利用离心力的
11. potential [pe'tenʃəl] n.（势）能，位（能），电势（位，压）
12. pedestrian [pi'destriən] a. 普通的，平凡的 n. 非专业人员
13. unimpressive [ˈʌnim'presiv] a. 给你印象不深的，平淡的，不令人信服的
14. inductive [in'dʌktiv] a. 引入的，导论的，诱导的，归纳的
15. awe [ɔ:] n.; v.（使）敬畏（畏惧）
16. cathedral [kə'θi:drəl] n.（一个教区内的）总教堂，大教堂，a.（像）大教堂的，权威的
17. metaphor ['metəfə] n. 隐喻，比喻
18. integrity [in'tegriti] n. 完整，完全，完善
19. grandeur ['grændʒə] n. 宏伟，壮观；伟大，崇高；富丽堂皇，豪华
20. implicitly [im'plicitli] ad. 含蓄地；无疑地，无保留地，绝对地
21. shaft [ʃɑ:ft] n.（传动，旋转）轴
22. reflux ['ri:flʌks] ad. 回流，倒流
23. condenser [kən'densə] n. 冷凝器
24. reboiler [ri'bɔilə] n. 再沸器
25. entropy ['entrəpi] n. 熵
26. immune [i'mju:n] a. 免除的，可避免的；不受影响的
27. spontaneous [spɔn'teinjəs] a. 自发的，自然的
28. heretofore ['hiətu'fɔ:] ad. 至今，到现在为止；在此以前
29. diminish [di'miniʃ] v. 减少，递减，削弱，由大变小

Notes

1. Lewis: Gilbert Newton Lewis, 1875～1946，美国物理化学家，提出了活度、逸度、离子强度等概念及共价键理论、广义酸碱理论等，1923 年与 M.Randall 合著《化学物质的热力学和自由能》一书。

Lesson 4
Chemical Reaction Engineering

Every industrial chemical process is designed to produce economically a desired product from a variety of starting materials through a succession of treatment steps. Fig. 4-1 shows a typical situation. The raw materials undergo a number of physical treatment steps to put them in the form in which they can be reacted chemically. They then pass through the reactor. The products of the reaction must then undergo further physical treatment—separations, purifications, etc.—for the final desired product to be obtained.

Design of equipment for the physical treatment steps is studied in the unit operations. Here we are concerned with the chemical treatment step of a process. Economically this may be an inconsequential unit, say a simple mixing tank. More often than not, however, the chemical treatment step is the heart of the process, the thing that makes or breaks the process economically.

Design of the reactor is no routine matter, and many alternatives can be proposed for a process. In searching for the optimum it is not just the cost of reactor that must be minimized. One design may have low reactor cost, but the materials leaving the unit may be such that their treatment requires much higher cost than alternative design. Hence, the economics of the over-all process must be considered.

Reactor design uses information, knowledge, and experience from a variety of areas—thermodynamics, chemical kinetics, fluid mechanics, heat transfer, mass transfer, and economics. Chemical reaction engineering is the synthesis of all these factors with the aim of properly designing a chemical reactor.

The design of chemical reactors is probably the one activity which is unique to chemical engineering, and it is probably this function more than anything else which justifies the existence of chemical engineering as a distinct branch of engineering.

In chemical reactor design there are two questions which must be answered:

(1) What change can we expect to occur?
(2) How fast will they take place?

The first question concerns thermodynamics, the second the various rate processes—chemical kinetics, heat transfer, etc. Putting these all together and trying to determine how these processes are interrelated can be an extremely difficult problem; hence we start with the simplest of situations and build up our analysis by considering additional factors until we are able to handle the more difficult problems.

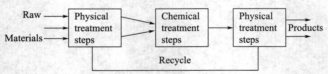

Fig.4-1 Typical chemical process

1. Thermodynamics

Thermodynamics gives two important pieces of information needed in design, the heat liberated or absorbed during reaction and the maximum possible extent of reaction.

Chemical reactions are invariably accompanied by the liberation or absorption of heat, the magnitude of which must be known for proper design. Consider the reaction

$$aA \longrightarrow rR + sS, \Delta H_r \quad \begin{cases} \text{positive, endothermic} \\ \text{negative, exothermic} \end{cases}$$

The heat of reaction at temperature T is the heat transferred from surroundings to the reacting system when a moles of A disappear to form r moles of R and s moles of S, with the system measured at the same temperature and pressure before and after reaction. With heats of reaction known or estimable from thermochemical data, the magnitude of the heat effects during reaction can be calculated.

Thermodynamics also allows calculation of the equilibrium constant K from the standard free energies of the reacting materials. With the equilibrium constant known, the expected maximum attainable yield of the products of reaction can be estimated.

2. Chemical Kinetics

Under appropriate conditions feed materials may be transformed into new and different materials which constitute different chemical species. If this occurs only by rearrangement or redistribution of the constituent atoms to form new molecules, we say that a chemical reaction has occurred. Chemistry is concerned with the study of such reactions. It studies the mode and mechanism of reactions, the physical and energy changes involved and the rate of formation of products.

It is the last-mentioned area of interest, chemical kinetics, which is of primary concern to us. Chemical kinetics searches for the factors that influence the rate of reaction. It measures this rate and proposes explanations for the values found. For the chemical engineer the kinetics of a reaction must be known if he is to satisfactorily design equipment to effect these reactions on a technical scale. Of course, if the reaction is rapid enough so that the system is essentially at equilibrium, design is very much simplified. Kinetic information is not needed, and thermodynamic information alone is sufficient.

3. Homogeneous and Heterogeneous Reactions

Homogeneous reactions are those in which the reactants, products, and any catalyst used form one continuous phase; gaseous or liquid. Homogeneous gas phase reactors will always be operated continuously; whereas liquid phase reactors may be batch or continuous. Tubular (pipe-line) reactors are normally used for homogeneous gas-phase reactions; for example, in the thermal cracking of petroleum crude oil fractions to ethylene, and the thermal decomposition of dichloroethane to vinyl chloride. Both tubular and stirred tank reactors are used for homogeneous liquid-phase reactions.

In a heterogeneous reaction two or more phases exist, and the overriding problem in the reactor design is to promote mass transfer between the phases. The possible combinations of phase are:

(i) Liquid-liquid: immiscible liquid phases; reactions such as the nitration of toluene or benzene with mixed acids, and emulsion polymerizations.

(ii) Liquid-solid: with one, or more, liquid phases in contact with a solid. The solid may be a reactant or catalyst.

(iii) Liquid-solid-gas; where the solid is normally a catalyst; such as in the hydrogenation of amines, using a slurry of platinum on activated carbon as a catalyst.

(iv) Gas-solid: where the solid may take part in the reaction or act as a catalyst. The reduction of iron ores in blast furnaces and the combustion of solid fuels are examples where the solid is a reactant.

(ⅴ) Gas-liquid: where the liquid may take part in the reaction or act as a catalyst.

4. Reactor Geometry (type)

The reactors used for established processes are usually complex designs which have been developed (have evolved) over a period of years to suit the requirements of the process, and are unique designs. However, it is convenient to classify reactor designs into the following broad categories.

Stirred Tank Reactors. Stirred tank (agitated) reactors consist of a tank fitted with a mechanical agitator and a cooling jacket or coils. They are operated as batch reactors or continuously. Several reactors may be used in series.

The stirred tank reactor can be considered the basic chemical reactor, modeling on a large scale the conventional laboratory flask. Tank sizes range from a few liters to several thousand liters. They are used for homogeneous and heterogeneous liquid-liquid and liquid-gas reactions; and for reactions that involve finely suspended solids, which are held by the agitation. As the degree of agitation is under the designer's control, stirred tank reactors are particularly suitable for reactions where good mass transfer or heat transfer is required.

When operated as a continuous process the composition in the reactor is constant and the same as the product stream, and, except for very rapid reactions, this will limit the conversion that can be obtained in one stage.

The power requirements for agitation will depend on the degree of agitation required and will range from about 0.2 kW/ms for moderate mixing to 2 kW/ms for intense mixing.

Tubular Reactor. Tubular reactors are generally used for gaseous reactions, but are also suitable for some liquid-phase reactions.

If high heat-transfer rates are required, small-diameter tubes are used to increase the surface area to volume ratio. Several tubes may be arranged in parallel, connected to a manifold or fitted into a tube sheet in a similar arrangement to a shell and tube heat exchanger. For high-temperature reactions the tubes may be arranged in a furnace.

Packed Bed Reactors. There are two basic types of packed-bed reactor: those in which the solid is a reactant, and those in which the solid is a catalyst. Many examples of the first type can be found in the extractive metallurgical industries.

In the chemical process industries the designer will normally be concerned with the second type: catalytic reactors. Industrial packed-bed catalytic reactors range in size from small tubes, a few centimeters diameter to large diameter packed beds. Packed-bed reactors are used for gas and gas-liquid reactions. Heat-transfer rates in large diameter packed beds are poor and where high heat-transfer rates are required fluidized beds should be considered.

Fluidized Bed Reactors. The essential features of a fluidized bed reactor is that the solids are held in suspension by the upward flow of the reacting fluid; this promotes high mass and heat-transfer rates and good mixing. The solids may be a catalyst; a reactant in fluidized combustion processes; or an inert powder, added to promote heat transfer. Though the principal advantage of a fluidized bed over a fixed bed is the higher heat-transfer rate, fluidized beds are also useful where it is necessary to transport large quantities of solids as part of the reaction processes, such as where catalysts are transferred to another vessel for regeneration.

Fluidization can only be used with relatively small sized particles, <300μm with gases.

A great deal of research and development work has been done on fluidized bed reactors in recent years, but the design and scale up of large diameter reactors is still an uncertain process and design methods are largely empirical.

Batch or Continuous Processing

In a batch process all the reagents are added at the commencement; the reaction proceeds the compositions changing with time, and the reaction is stopped and the product withdrawn when the required conversion has been reached. Batch processes are suitable for small-scale production and for processes where a range of different products, or grades is to be produced in the same equipment; for instance, pigments, dyestuffs and polymers.

In continuous processes the reactants are fed to the reactor and the products withdrawn continuously; the reactor operates under steady-state conditions. Continuous production will normally give lower production costs than batch production, but lacks the flexibility of batch production. Continuous reactors will usually be selected for large-scale production. Processes that do not fit the definition of batch or continuous are often referred to as semi-continuous or semi-batch. In a semi-batch reactor some of the reactants may be added, or some of the products withdrawn, as the reaction proceeds. A semi-continuous process can be one which is interrupted periodically for some purpose; for instance, for the regeneration of catalyst.

Selected from "Chemical Reaction Engineering, 2nd edition, O. Levenspiel, John Wiley & Sons, Inc., 1992"

New Words

1. endothermic [endəuˈθəmik] *a.* 吸热的
2. thermochemical [ˈθəːməuˈkemikəl] *a.* 热化学的
3. tubular reactor　管式反应器
4. stirred tank reactor　搅拌釜式反应器
5. amine [əˈmiːn] *n.* 胺
6. activated carbon　活性炭
7. blast furnace　高炉，鼓风炉
8. manifold [ˈmænifəuld] *n.* 总管，集气管，导管
9. tube sheet　管板
10. shell and tube heat exchanger　管壳式换热器，列管式换热器
11. extractive metallurgical　湿法冶金的
12. commencement [kəˈmensmənt] *n.* 开始，开端，开工

Exercise

1. Put the following into Chinese:
   thermal tubular reactor　　trickle bed reactor　　pursuit
   chemical vapor deposition　slurry reactor　　　　hydrodesulfrization
   ozonation　　　　　　　　chelation　　　　　　carbonylation
2. Put the following into English:
   吸热的　　　　放热的　　　　绝热的　　　　等温的
   连续的　　　　间歇的　　　　停留时间　　　返混
   均相的　　　　非均相的　　　管式反应器　　连续搅拌釜式反应器
   列管式反应器　湿法冶金的　　流体动力学　　多相反应器

Reading Material

Reactor Technology

Reactor technology comprises the underlying principles of chemical reaction engineering

(CRE) and the practices used in their application. Reactor designs evolve from the pursuit of new products and uses, higher conversion, more favorable reaction selectivity, reduced fixed and operating costs, intrinsically safe operation and environmentally acceptable processing. All the factors are interdependent and must be considered together, requirements for contacting reactants and removing products are a central focus in applying reactor technology; other factors usually are set by the original selection of the reacting system, intended levels of reactant conversion and product selectivity, and economic and environmental considerations. These issues should be taken into account when determining reaction kinetics from laboratory and bench-scale data, designing and operating pilot units, scaling up to large units, and ultimately in designing, operating, and improving industrial plant performance.

Most reactors have evolved from concentrated efforts focused on one type of reactor. Some processes have emerged from parallel developments using markedly different reactor types. In most cases, the reactor selected for laboratory study has become the reactor type used industrially because further development usually favors extending this technology. Following are illustrative examples of reactor usage, classified according to reactor type.

Batch Reactors. The batch reactor is frequently encountered in petrochemical, pharmaceutical, food, and mining processes, e.g., alkylation, emulsion polymerization, hydrocarbon fermentation, glycerolysis of fats, ozonation, and metal chelation. The processes often require achieving uniform dispersions of micrometer-sized drops and providing adequate exothermic heat removal especially where product quality is adversely affected by temperature.

Batch reactors are used in manufacturing plastic resins, eg, polyesters, phenolics, alkyds, urea-formaldehydes, acrylics, and furans. Such reactors generally are 6~40 m^3 baffled tanks, in which there are blades or impellers connected from above by long shafts, and heat is transferred either through jacketed walls or by internal coils.

Semibatch Reactors. Semibatch reactors are the most versatile of reactor types. Thermoplastic injection molds are semibatch reactors in which shaped plastic articles are produced from melts.

Reaction injection molding (RIM) circumvents the problems with injection molds and provides the technology for fabricating large articles, such as polyurethane automobile fenders and bumpers.

Continuous-Flow Stirred-Tank Reactors. The synthesis of p-tolualdehyde from toluene and carbon monoxide gas been carried out using CSTR equipment. p-Tolualdehyde (PTAL) is an intermediate in the manufacture of terephthalic acid. Hydrogen fluoride-boron trifluoride catalyzes the carbonylation of toluene to PTAL. In the industrial process, separate stirred tanks are used for each process step. Toluene and recycle HF and BF_3 come in contact in a CSTR to from a toluene-catalyst complex; the toluene complex reacts with CO to form the PTAL-catalyst complex in another CSTR. Once the complex decomposes to the product and the catalyst is regenerated, hydrated HF-BF_3 is processed in a separate vessel, because by-product water deactivates the catalyst and promotes corrosion.

Thermal Tubular Reactors. Tubular reactors have been widely used for low temperature, liquid-phase noncatalytic oxidation, e.g., butane to acetic acid and methyl ethyl ketone (MEK), p-xylene to terephthalic acid, cyclohexane to cyclohexanone and cyclohexanol, and n-alkanes to secondary alcohols, and high temperature pyrolysis, e.g., thermal cracking of petroleum feeds to olefins-particularly ethylene. Generally, conversion and selectivity to any given product are low for these oxidations. Because runaway branch-chain reactions are possible, heat dissipation must be assured and oxygen concentrations controlled. These considerations often favor the use of a series of tubular reactors in plug flow with some back-mixing in each reactor to maintain sufficient radical concentrations to propagate the reactions.

The epitaxy reactor is a specialized variant of the tubular reactor in which gas-phase precursors are produced and transported to a heated surface where thin crystalline films and gaseous by-products are produced by further reaction on the surface. Similar to this chemical vapor deposition (CVD) are physical vapor depositions (PVD) and molecular beam generated deposits. Reactor details are critical to assuring uniform, impurity-free deposits and numerous designs have evolved.

Bubble Columns. Bubble columns are finding increasing industrial application such as in ethylene dimerization, polymer manufacture, and liquid-phase oxidation. Bubble column processing offers the advantages of simplicity, favorable operating costs, and potentially superior product quality. Ethylene dimerization using homogeneous catalysts to produce 1-butene is favored compared to 1-butene produced by steam cracking, where butadiene must be separated and hydrogenated. Butadiene separation is complicated by the need to chemically remove the closely boiling isobutene, an impurity that later produces sticky polymer in some ethylene polymerizations.

Bubble columns in series have been used to establish the same effective mix of plug-flow and back-mixing behavior required for liquid-phase oxidation of cyclohexane, as obtained with staged reactors in series. Well-mixed behavior has been established with both liquid and air recycle.

Airlift Reactors. Airlift reactors are hydrodynamic variants of the bubble column in which the liquid or slurry circulates between two physically separated zones as a result of sparged gas in one zone inducing a density difference between the zones. The reactors have well-established uses in producing industrial and pharmaceutical chemicals, and in treating industrial and municipal wastes. Further applications stem from opportunities for waste minimization, e.g., converting ethylene and chlorine to dichloroethane. Important features in an airlift reactor design are the means used for ensuring liquid circulation, establishing high gas-liquid interfacial areas, and efficiently separating gas and liquid. Numerous configurations are in use including physically separated internal loops and draft tubes, and external loops. Also employed are vertical vessels with circulation induced by the downward injection of a bubbly liquid at velocities sufficient to force circulation.

Loop reactors are particularly suitable as bioreactors to produce, for example, single-cell protein. In this process, intense mixing and broth uniformity are essential, and must be accomplished without foaming, segregation of either buoyant or heavy components, or recourse to either chemical additives or mechanical agitation.

Spray Columns. Spray columns have diverse specialized uses in biotechnology processing, catalyst manufacture, and minimization of waste products. Spray columns are used in the production of milk powder, cheese, and other fermentation products for direct heat-induced conversion of proteins, microorganism and enzymes, thus affecting color, flavor, nutritive value, and biological safety. Incineration, pyrolysis, and partial oxidation are carried out in spray columns for the disposal of plastic wastes.

Tubular Fixed-Bed Reactors. Bundles of downflow reactor tubes filled with catalyst and surrounded by heat-transfer media are tubular fixed-bed reactors. Such reactors are used most notably in steam reforming and phthalic anhydride manufacture. Steam reforming is the reaction of light hydrocarbons, preferably natural gas or naphthas, with steam over a nickel-supported catalyst to form synthesis gas, which is primarily H_2 and CO with some CO_2 and CH_4.

Additional conversion to the primary products can be obtained by iron oxide-catalyzed water gas shift reactions, but these are carried out in large-diameter, fixed-bed reactors rather than in small-diameter tubes.

Fixed-Bed Reactors. Single-Phase Flow Fixed-bed reactors supplied with single-phase reactants are used extensively in the petrochemical industry, for ammonia synthesis, catalytic reforming, other hydroprocesses, e.g., hydrocracking and hydrodesulfurization, and oxidative dehydrogenation. The feeds in these processes are gases or vapors. The reactors generally are of large diameter, operate adiabatically, and often house multiple beds in individual pressure vessels. Bed geometries usually are determined by catalyst reactivity, so that the beds can be either tall and thin or short and squat. Variations of designs result from issues associated with the specific properties of the reactants or catalysts and differences in operating conditions. Either interbed cooling (or heating) or liquid quench additions are used to remove or supply reaction heat from the effluent of each bed. Multicomponent heterogeneous catalysts that require special activation treatments or unique handling requirements are used in most cases. Provisions must be made for restoring catalyst activity, either by replacement with fresh catalyst or by regeneration. Although reactants generally flow downward, air may be injected for catalyst regeneration at the bottom of the reactor.

Fixed-Bed Reactors. Multi-Phase Flow Flow regimes and contacting mechanisms in fixed-bed reactors that operate with mixtures of liquids and gases are totally different from those with single-phase feeds. Nevertheless, mixed-phase, downflow fixed-bed reactor designs are extensions of single-phase, fixed bed hydroprocessing technology and outwardly resemble such reactors. The most generally used mixed-phase reactor is the trickle bed. Special distributors are used to uniformly feed the two-phase mixtures. Hydrodesulfurization, hydrocracking, hydrogenation, and oxidative dehydrogenation are carried out with high boiling feeds in such reactors. Pressures generally are higher than in their single-phase flow counterparts. Though some of these reactors may operate in the pulsed-flow regime, these reactors retain their trickle bed name because the same configurations are used. Furthermore, reactor instrumentation may not be suited for recording the pulsation, thus the flow regime would not be noted. Where required, the temperature rise in trickle-bed reactors is controlled as with single-phase reactors. Generally, liquid quench systems are preferred.

Increased global urbanization provides the impetus to exploit improvements in feed distribution and catalyst utilization to achieve very low (<500 ppm) sulfur levels in existing hydrodesulfurization reactors for a wide range of feeds. High boiling feeds that have been successfully processed include virgin and cracked heavy gas oils and residue. Such feeds contain metals, which are hydrodemetallized and removed by deposition within the catalyst porous structures. Catalysts that are low in cost relative to noble metal supported catalysts slowly deactivate and are replaced after one to three years of use, depending on the severity of service. Vapor-liquid equilibria play an important role in catalyst deactivation. Moving trickle beds are also used in these applications. Catalysts can be fed as an oil slurry to the top and then flow downward through a series of beds. The conical bottom of each bed is designed to assure plug flow of catalyst.

Selected from "Encyclopedia of Chemical Technology, Vol. 20, R.E.Kirk and D.F. Othmer, Interscience, 3rd edition 1996"

New Words

1. pursuit [pəˈsjuːt]   *n.* 追求，从事，研究
2. glycerolysis of fats   脂肪甘油水解
3. ozonation   *n.* 臭氧化作用
4. chelation [kiˈleiʃən]   *n.* 螯合作用

5. phenolics [fi'nɔliks] n. 酚醛塑料（树脂）
6. furan ['fjuəræn] n. 呋喃
7. fender ['fendə] n. 挡泥板，防护板
8. bumper ['bʌmpə] n. 保险杠
9. *p*-tolualdehyde n. 对甲苯甲醛
10. CSTR=continuously stirred tank reactor 连续搅拌釜反应器
11. carbonylation n. 羰化作用
12. deactivate [diːˈæktiveit] vt. 减活，去活化，钝化
13. thermal tubular reactor 热管反应器
14. ketone ['kiːtəun] n. (甲)酮
15. cyclohexanone [ˌsaikləu'heksənəun] n. 环己酮
16. cyclohexanol [saikləu'heksənɔl] n. 环己醇
17. secondary ['sekəndəri] 仲 [指 CH_3—$CH(CH_3)$-型支链烃，或指二元胺及 R_2CHOH 型的醇]
18. back-mixing 返混
19. epitaxy [epi'tæksi] n. 晶体取向生长
20. chemical vapor deposition 化学气相淀积
21. physical vapor deposition 物理气相淀积
22. molecular beam 分子束
23. molecular beam generated deposit 分子束淀积
24. bubble column 鼓泡塔
25. dimerization [daiməraiˈzeiʃən] n. 二聚（作用）
26. isobutene [aisəu'bjuːtiːn] n. 异丁烯
27. airlift reaction 升气式反应器
28. hydrodynamics ['haidrəudaiˈnæmiks] n. 流体动力学
29. municipal [mjuːˈnisipl] a. 市政的，城市的
30. loop reactor 环路反应器
31. segregation [segri'geiʃən] n. 分离，分凝，分开
32. incineration [insinəˈreiʃən] n. 焚化，灰化，煅烧
33. tubular fixed-bed reaction 管式固定床反应器
34. phthalic anhydride 邻苯二甲酸酐
35. water gas 水煤气
36. hydroprocess 加氢过程
37. hydrodesulfurization 加氢脱硫过程
38. mixed-phase reactor 多相反应器
39. counterpart ['kauntəpɑːt] n. 对应物，配对物，对方；一对东西中之一，副本
40. trickle bed reactor 滴流床反应器
41. impetus ['impitəs] n. (推)动力，促进，动量，冲量
42. sulfur=sulphur
43. virgin oil 直馏油
44. hydrodementallize v. 加氢脱金属
45. moving trickle bed reactor 移动滴流床反应器

Lesson 5
Chlor-Alkali and Related Processes

Historically the bulk chemical industry was built on chlor-alkali and related processes. The segment is normally taken to include the production of chlorine gas, caustic soda (sodium hydroxide), soda-ash (derivatives of sodium carbonate in various forms) and, for convenience, lime-based products.

Soda-ash and sodium hydroxide have competed with each other as the major source of alkali ever since viable processes were discovered for both. The peculiar economics of electrolytic processes mean that you have to make chlorine and caustic soda together in a fixed ratio whatever the relative demand for the two totally different types of product, and this causes swings in the price of caustic soda which can render soda-ash more or less favorable as an alkali.

Both chlorine/caustic soda and soda-ash production are dependent on cheap readily available supplies of raw materials. Chlorine/caustic soda requires a ready supply of cheap brine and electricity, soda-ash requires brine, limestone and lots of energy. Soda-ash plants are only profitable if their raw materials do not have to be transported far. The availability of such supplies is a major factor in the location of many of the chemical industry's great complexes.

1. Lime-Based Products

One of the key raw materials is lime. Limestone consists mostly of calcium carbonate ($CaCO_3$) laid down over geological time by various marine organisms. High quality limestones are often good enough to be used directly as calcium carbonate in further reactions. Limestone is usually mined in vast open-cast quarries, many of which will also carry out some processing of the materials.

The two key products derived from limestone are quicklime (CaO) and slaked lime [$Ca(OH)_2$]. Quicklime is manufactured by the thermal decomposition (1200~1500℃) of limestone according to the equation:

$$CaCO_3 \longrightarrow CaO + CO_2$$

Typically limestone is crushed and fed into the higher end of a sloping rotating kiln[1] where the decomposition takes place and quicklime is recovered from the end. Most frequently, however, the quicklime is not isolated for further reactions, rather, other compounds are fed in with the lime to give final products at the low end of the kiln. For example, alumina, iron ore and sand can be fed in to give Portland cement[2]. Soda-ash manufacture often adds coke to the limestone which burns to give extra carbon dioxide needed for soda-ash manufacture. Slaked lime which is more convenient to handle than quicklime—is manufactured by reacting quicklime with water.

About 40% of the output of the lime industry goes into steel-making, where it is used to react with the refractory silica present in iron ore to give a fluid slag which floats to the surface and is easily separated from the liquid metal. Smaller, but still significant, amounts are used in chemical manufacture, pollution control and water treatment. The most important chemical derived from lime is soda-ash.

2. Soda Ash

The Solvay Process[3,4]. The process, which was perfected by Ernest Solvay in 1865, is based on the precipitation of $NaHCO_3$ when an ammoniated solution of salt is carbonated with CO_2 from a coke-fired lime kiln. The $NaHCO_3$ is filtered, dried, and calcined to Na_2CO_3. The filtered ammonium chloride process liquor is made alkaline with slaked lime and the ammonia is distilled out for recycle to the front end of the process. The resultant calcium chloride is a waste or by-product stream.

For a simple basic product the Solvay process appears exceedingly complicated. The basic principle of the reaction is to take salt (NaCl) and calcium carbonate ($CaCO_3$) as inputs and to produce calcium chloride and sodium carbonate as outputs. However, the reactions occurring between input and output are not remotely obvious and involve the use of ammonia and calcium hydroxide as intermediate compounds.

The essential principle is that, by carefully controlling the concentration of the components (especially ammonia and salt), sodium bicarbonate can be precipitated from solutions containing salt, carbon dioxide and ammonia. The key to making the process work is controlling the strength of the solutions and the rates of crystallization.

The essential steps of the process are as follows. Ammonia is absorbed in an ammonia absorber into brine which has previously been purified to reduce the amount of calcium and magnesium ions (which tend to precipitate during the process in all the wrong places, blocking pipe-work). The solution (nominally containing sodium chloride and ammonium hydroxide) is then passed down a tower where it absorbs carbon dioxide (passing up the tower) to form ammonium carbonate at first and later ammonium bicarbonate. By the next stage of the plant sodium chloride and ammonium bicarbonate have metathesised to sodium bicarbonate (which precipitates) and ammonium chloride. Filtration separates the solid bicarbonate from the remaining solution. The bicarbonate is passed to a rotary dryer where it loses water and carbon dioxide to give a fluffy crystalline mass known as light soda-ash which is mostly sodium carbonate. The fluffy mass is light because the original crystal shape is retained on the loss of carbon dioxide, leaving many voids. It is usually more convenient to make a more dense material and this is achieved by adding water (which causes recrystallization in a denser form) and further drying.

It is debatable whether the actual chemistry given above is a good description of the process, but it certainly aids understanding. For a detailed understanding, a great deal needs to be known about solubility products of multicomponent systems. The important thing to know is that the system is complex and requires careful control at all parts of the process in order for it to operate effectively.

One disadvantage of the process is the amount of calcium chloride produced. Far more is produced than can be used, so much of the production is simply dumped (it is not a particularly noxious or nasty product). It would be advantageous to use all the input material in this process, for example producing hydrogen chloride from the chloride.

Use of Soda Ash. Of all soda ash, 50% is sold to the glassmaking industry as it is a primary raw material for glass manufacture. The fortunes of the industry are therefore strongly tied to glass demand. Soda ash also competes directly with sodium hydroxide as an alkali in many chemical processes. Sodium silicates are another important class of chemicals derived from soda ash by reaction with silica at 1200~1400℃. Silica-gel is a fine sodium silicate with a large surface area and is used in catalysts, chromatography and as a partial phosphate replacement in detergents and soaps.

3. Electrolytic Processes for Chlorine/Caustic Soda

Introduction. Both chlorine and caustic soda have, at various times in the history of the chemical industry, been greatly in demand, but unfortunately for operators of electrochemical plants, not always at the same time. Chlorine has been valued as a bleach, or a raw material for the production of bleaching powder, as a disinfectant in water supplies and as a raw material for plastics and solvents manufacture. Caustic soda has been used in the production of soda ash, soap, textiles, and as a very important raw material in an incredible variety of chemical processes.

All the electrolytic processes have in common the electrolysis of salt to give chlorine and sodium hydroxide. The vast majority of production electrolyses a solution of salt, but there are some significant plants that electrolyze molten salt to give liquid sodium and chlorine. These are used by industries that need the liquid sodium, mainly in the production of tetra-alkyl lead petroleum additives, though the petroleum additive companies are diversifying and other uses may appear. There are essentially three different types of cell used for aqueous electrolysis: mercury cells, diaphragm cells and membrane cells. Membrane cells are really the only technology that is viable for new capacity in modern plants, but a large amount of old capacity still exists and many companies have not found it economical to replace even their mercury cells, despite the environmental implications.

All electrolytic reactions are based on the idea of using electrons as a reagent in chemical reactions. The basic reactions of brine electrolysis can be written as follows:

$$\text{Anode } 2Cl^- - 2e^- \longrightarrow Cl_2$$
$$\text{Cathode } 2H_2O + 2e^- \longrightarrow H_2 + 2OH^-$$

The overall reaction is:

$$2Na^+ + 2Cl^- + 2H_2O \longrightarrow NaOH + Cl_2 + H_2$$

This reaction has a positive free energy (ΔG=421.7kJ/mol at 25℃) and needs to be driven uphill by electricity.

Like many basic chemical processes, though the reaction appears to be gloriously simple, there are some significant complications. For a start, the reaction products need to be kept apart: hydrogen and chlorine will react explosively if they are allowed to mix. Chlorine reacts with hydroxide to give hypochlorous acid (HOCl) and chloride (both wasting product and creating by-products). The hypochlorous acid and hypochlorite (ClO^-) in turn react to give chlorate (ClO_3^-), protons and more chloride. Hydroxide reacts at the anode to form oxygen, which can contaminate the chlorine. All the reactions reduce efficiency and/or create difficult separation of contamination problems that need to be sorted out before any products can be sold. The key to understanding the various types of process used for the electrolysis is the way they separate the reaction products. There are basically three types of electrolytic cell for brine electrolysis, though there are many variations of detail among the cells from different manufactures.

4. The Uses of Chlorine and Sodium Hydroxide

Sodium hydroxide has so many chemical uses that it is difficult to classify them conveniently. One of the largest uses is for paper-making, where the treatment of wood requires a strong alkali. In some countries this consumes 20% of production. Another 20% is consumed in the manufacture of inorganic chemicals such as sodium hypochlorite (the bleach and disinfectant). Various organic syntheses consume about another fifth of the production. The production of alumina and soap uses smaller amounts.

Chlorine is widely used in a variety of other products. About a quarter of all production world-wide goes into vinyl chloride, the monomer for making PVC. Between a quarter and a half goes into a variety of other products. Depending on the country, up to 10% goes into water purification. Up to 20% goes into the production of solvents (methylchloroform, trichloroethene, etc.) though many of these are being phased out because of the Montreal Protocol. About 10% world-wide goes into the production of inorganic chlorine-containing compounds. A very significant use in some countries is for the bleaching of wood pulp, though this is another use coming under environmental pressure.

Selected from "An Introduction to Industrial Chemistry, 2nd Edition, C.A. Heaton, Blackie & Son Ltd., 1997"

New Words

1. segment ['segmənt]   *n.* 部分；切片
2. soda ash   *n.* 纯碱，无水碳酸钠，苏打灰，碱灰
3. peculiar [pi'kju:ljə]   *a.* 特有的，独特的，特殊的；奇怪的
4. electrolytic [i'lektrəu'litik]   *a.* 电解的，电解质的
5. open-cast   *a.; ad.* 露天开采的（地）
6. quicklime ['qwiklaim]   *n.* 生石灰，氧化钙
7. slakedlime   *n.* 熟石灰，消石灰
8. kiln [kiln，kil]   *n.* 窑，炉 *v.* 窑烧
9. decomposition [di:kəmpə'ziʃən] *n.* 分解，离解
10. alumina [ə'lu:minə]   *n.* 矾土，氧化铝
11. Protland cement   硅酸盐水泥，波兰特水泥，普通水泥
12. refractory [ri'fræktəri]   *a.* 难熔的，耐火的； *n.* 耐火材料
13. slag [slæg]   *n.* （炉，熔，矿）渣
14. the Solvay process   索尔维法
15. ammoniated [ə'məunieitid]   *a.* 充氨的，含氨的
16. calcine ['kælsin]   *v.; n.* 煅烧，烧成（灰）
17. alkaline [ælkəlain]   *n.* 碱性   *a.* 强碱的
18. distil(l) [dis'til]   *vt.* 蒸馏，用蒸馏法提取；提取……的精华
19. bicarbonate [bai'kɑ:bənit]   *n.* 碳酸氢盐，酸式碳酸盐
20. crystallization [kristəlai'zeiʃən]   *n.* 结晶（作用，过程）
21. metathesis [me'tæθəsis]   *n.* 复分解（作用），置换（作用）
22. filtration [fil'treiʃən]   *n.* 过滤
23. rotary dryer   旋转干燥器
24. fluffy ['flʌfi]   *a.* 蓬松的，松软的
25. crystalline ['kristəlain]   *a.* 结晶的，结晶状的；水晶的
26. void [vɔid]   *n.* 空隙，空隙率；空间，空位
27. solubility [sɔlju'biliti]   *n.* 溶解度，溶解性
28. solubility product   溶度积
29. noxious ['nɔkʃəs]   *a.* 有毒的，有害的，不卫生的
30. nasty ['nɑ:sti]   *a.* 难处理的，极脏的，（气味）令人作呕的
31. silicate ['silikit]   *n.* 硅酸盐（酯）
32. silica-gel   （氧化）硅胶
33. chromatography [,krəumə'tɔgrəfi]   *n.* 色谱（法，学），色层法
34. electrochemical [i'lektrə'kemikəl]   *a.* 电化学的
35. disinfectant [disin'fektənt]   *n.* 消毒剂，杀菌剂
36. incredible [in'kredəbl]   *a.* 难以置信的，不可思议的，惊人的
37. electrolysis [ilek'trɔlisis]   *n.* 电解法，电解作用，电分析
38. electrolyse [i'lektrəlaiz]   *vt.* 电解（=electrolyze）
39. tetra-alkyl lead   四烷基铅
40. mercury ['mə:kjuri]   *n.* 汞，水银 (Hg)
41. diaphragm ['daiəfræm]   *n.* 隔膜，隔膜
42. membrane ['membrein]   *n.* 膜，膜片，隔膜
43. anode ['ænəud]   *n.* 阳极，正极
44. cathode ['kæθəud]   *n.* 阴极，负极

45. hypochlorous acid 次氯酸
46. hypochlorite [haipə'klɔːrait] *n.* 次氯酸盐
47. chlorate ['klɔːrit] *n.* 氯酸盐
48. proton ['prəutɔn] *n.* 质子
49. contaminate [kən'tæmineit] *vt.* 污染，弄脏，毒害
50. methylchloroform *n.* 三氯乙烷，甲基氯仿
51. trichloroethene 三氯乙烯
52. pulp [pʌlp] *n.* 浆状物，纸浆；矿浆

Notes

1. kiln: 窑，指高温下（一般大于 800℃）通过焙烧（roasting）过程烧制产品的热工设备。roasting: 焙烧，固体物料在高温不发生熔融的条件下进行的化学反应，可以有氧化、热解、还原、卤化等。其中不加添加剂的焙烧，也称为煅烧，如石灰石化学加工制成氧化钙，同时制得二氧化碳气体。

2. portland cement: 硅酸盐水泥，一类以高碱性硅酸盐为主要化合物的水硬性水泥的总称，因其凝结硬固后的外观、颜色与早期英国用于建筑的优质波特兰石头相似，故西方国家统称为波特兰水泥。

3. the Solvay process: 索尔维法，又称氨碱法，由比利时人 Ernest Solvay 提出，是纯碱生产的最主要方法。先将原盐溶化成饱和盐水，除去杂质，然后吸收氨制成氨盐水，再进行碳化得碳酸氢钠，过滤后煅烧而得纯碱。

4. Ernest Solvay: 1838~1922，比利时工业化学家，1861 年，他在煤气厂从事稀氨水的浓缩工作时，在用盐水吸收氨和二氧化碳的实验中得到碳酸氢钠，同年，他获得了用食盐、氨和二氧化碳制取碳酸钠的工业生产方法的专利，称为索尔维法，又称氨碱法。

Exercise

1. Put the following into Chinese:

oleum	mercury	soda ash	metathesis
PVC	alkaline	desulphurization	membrane
carbonate	caustic sodium	proton	polytetrafluoroethylene

2. Put the following into English:

电解的	分解	氯化物	还原
催化剂	氧化反应	动力学	沉淀
钙	镁	树脂	表面活性剂

Reading Material

Sulphuric Acid

1. Introduction

Sulphuric acid is the chemical that is produced in the largest tonnage. Such is its importance as a raw material for other processes that its production was, until recently, considered a reliable indicator of a country's industrial output and the level of its industrial development. It is still used in an incredible variety of different processes and remains one of the most important chemicals produced.

2. Raw Materials

The raw material for sulphuric acid production is elemental sulphur, which can be obtained from several sources. Almost all the elemental sulphur produced is used in the manufacture of sulphuric acid. The biggist source used to be the direct mining of underground deposits of sulphur by the Frasch process[1]. This involves injection of superheated water and air into the deposits of sulphur via drilling. The resulting aerated liquid sulphur-water-air mixture is buoyant enough to rise to the drill head where it can be separated into its components, and pure sulphur is recovered.

The petrochemical industry now provides more sulphur (from the desulfurisation of oil and gas) than the Frasch process. Some petrochemical deposits contain large quantities of sulphur-containing compounds (25% in some Russian deposits) which must be removed to avoid poisoning the cracking catalysts or indeed the public (though the residual sulphur in some petrol is enough to cause bad hydrogen sulfide smells from some catalytic converters). The process that removes the sulphur creates hydrogen sulfide which is easy to separate from oil and gas and is easily converted to elemental sulphur.

Some sulphur is also produced as a by-product of metal extraction. Many sulfide-containing metal ores are burned as part of the extraction process, giving off sulphur dioxide, which can be recovered and used in the sulphuric acid industry.

One untapped source of sulphur is the coal used in electricity generation. More sulphur is emitted from power stations than is used by the chemical industry. Most of this ends up in the atmosphere where it causes acid rain, though power stations are increasingly having to scrub their flue gases. Unfortunately, there is at present no convenient way to recover this sulphur in a useable form.

3. The Manufacturing Process

The production of sulphuric acid has three stages:

(1) The burning of sulphur in air to give sulphur dioxide

$$S + O_2 \longrightarrow SO_2$$

(2) The reaction of sulphur dioxide and oxygen to give sulphur trioxide

$$2SO_2 + O_2 \longrightarrow 2SO_3$$

(3) The absorption of sulphur trioxide in water to give sulphuric acid.

The first stage is simple, with few of the all too common complications that beset many industrially important "simple" reactions. Molten sulphur is sprayed into a furnace in a current of dry air at about 1000℃ to produce a gas stream containing about 10% sulphur dioxide. The stream is cooled in a boiler where the energy of the exothermic reaction can be extracted and the temperature brought down to 420℃.

The second stage is the key to the process. The direct reaction between sulphur dioxide and oxygen to give sulphur trioxide is slow and requires a catalyst. In the old lead chamber process[2] nitrogen dioxide (NO_2) was used for the oxidation though in practice mixtures of nitrogen oxides were used. This process has now been completely superseded by the contact process, which speeds the direct reaction via a solid-state catalyst of vanadium pentoxide (V_2O_5). The catalyst is normally absorbed on an inert silicate support and lasts for about 20 years.

The reaction between oxygen and sulphur dioxide is exothermic so the equilibrium favors the product at lower temperature. So much heat is produced by the reaction that it is difficult to achieve good yields in a single-bed reactor: the reactor warms up so much that the reaction goes into reverse. Most plants therefore used several reactor stages (with coolers between the reactors) so that the heat produced does not drive the reaction backwards. The first reactor chamber converts the mixture of oxygen and sulphur dioxide into a stream about 60%~65% converted to products at about 600℃. It is then cooled to about 400℃ and passed into the next layer of catalyst, and so on. After three layers 95%~96% of the starting materials have been converted to products (this is near the maximum possible conversion unless sulphur trioxide is removed).

The gas mixture can then be passed into the initial absorption tower where some sulphur trioxide is removed. The remaining gas mixture is then reheated and passed back into a fourth converter which enables overall conversion of up to 99.7% to be achieved.

The final stage involves passing the gas mixture into the final absorption tower where the sulphur trioxide is hydrated with water to give, at the end of the tower, a 98%~99% acid solution. If excess sulphur trioxide is used the mixture consists of, effectively, sulphur trioxide dissolved in pure sulphuric acid this is known as oleum.

4. Uses of Sulphuric Acid

A large amount (66% of total production in the peak year) of sulphuric acid is used in the manufacture of phosphoric acid for fertilizer (phosphate rock plus acid gives impure phosphoric acid), though this use is now declining as demand for fertilizers declines. Some more goes into production of ammonium sulphate, a low-grade fertilizer.

More interesting are some of the more speciality uses of the compound, which are less significant in volume but are possibly more important for the rest of the industry. There are many important large-scale industrial syntheses that use sulphuric acid, for example the manufacture of ethanol from ethene, one of the routes to titanium dioxide (a white pigment important to the paint industry), the production of hydrofluoric acid from calcium fluoride (the ultimate source of about 70% of all the fluorine in fluorinated compounds), the production of aluminum sulphate (an important water-treatment chemical), the production of sulphonated surfactants (detergents and many other applications). Many uses are based on the fact that sulphuric aicd represents a cheap source of protons. For example, the manufacture of hydrofluoric acid from fluoride-containing minerals involves mixing them with concentrated sulphuric acid in a rotating kiln the fluoride is protonated to hydrogen fluoride by the acid and the hydrogen fluoride is given off as a gas.

Selected from "An Introduction to Industrial Chemistry, 2nd Edition, C.A. Heaton, Blackie & Son Ltd., 1997"

New Words

1. Frash process   地下熔融法
2. aerate [ˈɛəreit] vt. 充气，鼓气，通风，鼓风
3. buoyant [ˈbɔiənt] a. 有浮力的，能浮的，易浮的
4. desulphurization [diːsʌlfəraiˈzeiʃən] n. 脱硫，除硫
5. sulfide [ˈsʌlfaid] n. 硫化物
6. untapped a. 未利用的，未开发的
7. flue [fluː] n. 烟道，风道
8. beset [biˈset] vt. 包围，缠绕，为……所苦
9. exothermic [eksəuˈθəːmik] a. 放热的
10. lead chamber process   铅室法
11. supersede [sjuːpəˈsiːd] vt. 代替，取代，废弃
12. vanadium [vəˈneidjəm] n. 钒 (V)
13. inert [iˈnəːt] a. 惰性的，不活泼的；惯性的
14. equilibrium [iːkwiˈlibriəm] n. 平衡
15. absorption [əbˈsɔːpʃən] n. 吸收（作用）
16. oleum n. 发烟硫酸
17. hydrofluoric [haidrəufluˈɔrik] a. 氟化氢的，氢氟酸的

18. fluorine ['fluəri:n]   n. 氟 (F)
19. fluorinated ['fluərineitid]   a. 氟化的
20. surfactant [sə:'fæktənt]   n. 表面活性剂
21. protonate   v. 使质子化

Notes

1. Frasch process: 地下熔融法，又称弗拉施法，由美国人 Herman Frasch (1851—1914) 于 1894 年发明。先在矿区地表钻孔至含硫层，然后插入直径各为 20cm、10cm、2.5cm 的三层同心套管，向外管中压入 160~165℃ 的过热水，使地下硫黄熔化，再向直径 2.5cm 的管内通入热的压缩空气，使液硫从 10cm 的管上升压出，液硫经脱气后固化，或以液态储运，产品纯度可达 99.7%~99.8%。

2. lead chamber process: 铅室法，利用高级氮氧化物（主要是三氧化二氮使二氧化硫氧化并生成硫酸），因以铅制的方形空室为主要设备而得名，是硫酸工业发展史上最古老的工业生产方法。

Lesson 6
Ammonia

Dinitrogen makes up more than three-quarters of the air we breathe, but it is not readily available for further chemical use. Biological transformation of nitrogen into useful chemicals is embarrassing for the chemical industry, since all the effort of all the industry's technologists has been unable to find an easy alternative to this. Leguminous plants can take nitrogen from the air and convert it into ammonia and ammonium-containing products at atmosphere pressure and ambient temperature; despite a hundred years of effort, the chemical industry still needs high temperatures and pressures of hundreds of atmospheres to do the same job. Indeed, until the invention of the Haber process, all nitrogen-containing chemicals came from mineral sources ultimately derived from biological activity.

Essentially all the nitrogen in manufactured chemicals comes from ammonia derived from the Haber-based process. So much ammonia is made (more molecules than any other compound, though because it is a light molecule greater weights of other products are produced), and so energy-intensive is the process, that ammonia production alone was estimated to use 3% of the World's energy supply in the mid-1980s.

The Haber Process For Ammonia Synthesis

Introduction. All methods for making ammonia are basically fine-tuned versions of the process developed by Haber, Nernst and Bosch in Germany just before the First World War.

$$N_2 + 3H_2 \rightleftharpoons 2NH_3$$

In principle the reaction between hydrogen and nitrogen is easy; it is exothermic and the equilibrium lies to the right at low temperatures. Unfortunately, nature has bestowed dinitrogen with an inconveniently strong triple bond, enabling the molecule to thumb its noise at thermodynamics. In scientific terms the molecule is kinetically inert, and rather severe reaction conditions are necessary to get reactions to proceed at a respectable rate. A major source of "fixed" (meaning, paradoxically, "usefully reactive") nitrogen in nature is lightning, where the intense heat is sufficient to create nitrogen oxides from nitrogen and oxygen.

To get a respectable yield of ammonia in a chemical plant we need to use a catalyst. What Haber discovered—and it won him a Noble prize—was that some iron compounds were acceptable catalysts. Even with such catalysts extreme pressures (up to 600 atmospheres in early processes) and temperatures (perhaps 400℃) are necessary.

Pressure drives the equilibrium forward, as four molecules of gas are being transformed into two. Higher temperatures, however, drive the equilibrium the wrong way, though they do make the reaction faster, chosen conditions must be a compromise that gives an acceptable conversion at a reasonable speed. The precise choice will depend on other economic factors and the details of the catalyst. Modern plants have tended to operate at lower pressures and higher temperatures (recycling unconverted material) than the nearer-ideal early plants, since the capital and energy costs have become more significant.

Biological fixation also uses a catalyst which contains molybdenum (or vanadium) and iron embedded in a very large protein, the detailed structure of which eluded chemists until late 1992. How it works is still not understood in detail.

Raw Materials. The process requires several inputs: energy, nitrogen and hydrogen, Nitrogen is easy to extract from air, but hydrogen is another problem. Originally it was derived from coal via coke which can be used as a raw material (basically a source of carbon) in steam reforming, where steam is reacted with carbon to give hydrogen, carbon monoxide and carbon dioxide. Now natural gas (mainly methane) is used instead, though other hydrocarbons from oil can also be used. Ammonia plants always include hydrogen-producing plants linked directly to the production of ammonia.

Prior to reforming reactions, sulphur-containing compounds must be removed from the hydrocarbon feedstock as they poison both the reforming catalysts and the Haber catalysts. The first desulphurization stage involves a cobalt-molybdenum catalyst, which hydrogenates all sulphur-containing compounds to hydrogen sulfide. This can then be removed by reaction with zinc oxide (to give zinc sulfide and water).

The major reforming reactions are typified by the following reactions of methane (which occur over nickel-based catalysts at about 750℃):

$$CH_4 + H_2O \longrightarrow CO + 3H_2$$
$$\text{Synthesis gas}$$
$$CH_4 + 2H_2O \longrightarrow CO_2 + 4H_2$$

Other hydrocarbons undergo similar reactions.

In the secondary reformers, air is injected into the gas stream at about 1100℃. In addition to the other reactions occurring, the oxygen in the air reacts with hydrogen to give water, leaving a mixture with close to the ideal 3:1 ratio of hydrogen to nitrogen with no contaminating oxygen. Further reactions, however, are necessary to convert more of the carbon monoxide into hydrogen and carbon dioxide via the shift reaction:

$$CO + H_2O \longrightarrow CO_2 + H_2$$

This reaction is carried out at lower temperatures and in two stages (400℃ with an iron catalyst and 220℃ with a copper catalyst) to ensure that conversion is as complete as possible.

In the next stage, carbon dioxide must be removed from the gas mixture, and this is accomplished by reacting the acidic gas with an alkaline solution such as potassium hydroxide and/or mono-or di-ethanolamine.

By this stage there is still too much contamination of the hydrogen-nitrogen mixture by carbon monoxide (which poisons the Haber catalysts), and another step is needed to get the amount of CO down to ppm levels. This step is called methanation and involves the reaction of CO and hydrogen to give methane (i.e. the reverse of some of the reforming steps). The reaction operates at about 325℃ and uses a nickel catalyst.

Now the synthesis gas mixture is ready to go into a Haber reaction.

Ammonia Production. The common features of all the different varieties of ammonia plant are that the synthesis gas mixture is heated, compressed and passed into a reactor containing a catalyst. The essential equation for the reaction is simple:

$$N_2 + 3H_2 \rightleftharpoons 2NH_3$$

What industry needs to achieve in the process is an acceptable combination of reaction speed and reaction yield. Different compromises have been sought at different times and in different economic circumstances. Early plants plumped for very high pressure (to get the yield up in a one-pass reactor), but many of the most modern plants have accepted much lower one-pass yields at lower pressures and have also opted for lower temperatures to conserve energy. In order to ensure the maximum yield in the reactor the synthesis gas is usually cooled as it reaches equilibrium. This can be done by the use of heat exchangers or by the injection of cool gas into the reactors at an appropriate point. The effect of this is to freeze the reaction as near to equilibrium as possible. Since the reaction is exothermic (and the equilibrium is less favorable for ammonia synthesis at higher temperatures) the heat must be carefully controlled in this way to achieve good yields.

The output from the Haber stage will consist of a mixture of ammonia and synthesis gas so the next stage needs to be the separation of the two so that the synthesis gas can be recycled. This is normally accomplished by condensing the ammonia (which is a good deal less volatile than the other components, ammonia boils at about $-40\,^\circ C$).

Use of Ammonia. The major use of ammonia is not for the production of nitrogen-containing chemicals for further industry use, but for fertilizers such as urea or ammonium nitrates and phosphates. Fertilizers consume 80% of all the ammonia produced. In the USA in 1991, for example, the following ammonia-derived products were consumed, mostly for fertilizers (amounts in millions of tonnes): urea (4.2); ammonium sulphate (2.2); ammonium nitrate (2.6); diammonium hydrogen phosphate (13.5).

Chemical uses of ammonia are varied. The Solvay process for the manufacture of soda ash uses ammonia, though it does not appear in the final product since it is recycled. A wide variety of processes take in ammonia directly, including the production of cyanides and aromatic nitrogen-containing compounds such as pyridine. The nitrogen in many polymers (such as nylon or acrylics) can be traced back to ammonia, often via nitriles or hydrogen cyanide. Most other processes use nitric acid or salts derived from it as their source of nitrogen. Ammonium nitrate, used as a nitrogen-rich fertilizer, also finds a major use as a bulk explosive.

Selected from "An Introduction to Idustrial Chemistry, 2nd Edition, C. A. Heaton, Blackie & Son Ltd., 1997"

New Words

1. dinitrogen [daiˈnaitrədʒən] *n.* 分子氮，二氮
2. leguminous [leˈgjuːminəs] *a.* 豆科的，似豆科植物的
3. ambient [ˈæmbiənt] *a.* 周围的，包围着的
4. bestow [beˈstəu] *vt.* 把……赠予（给）
5. thumb one's noise (at) （对……）作蔑视的手势
6. thermodynamics [ˌθəːməudaiˈnæmiks] *n.* 热力学
7. paradoxically [ˌpærəˈdɔksikəli] *ad.* 似非而可能是，自相矛盾地，荒谬地
8. molybdenum [mɔˈlibdinəm] *n.* 钼 (Mo)
9. embed [imˈbed] *vt.* 把……嵌入；栽种
10. protein [ˈprəutiːn] *n.* 蛋白质
11. elude [iˈluːd] *vt.* 使困惑，难倒
12. cobalt [ˈkəubɔːlt] *n.* 钴 (Co)
13. hydrogenate [haiˈdrɔdʒəneit] *vt.* 使与氢化合，使氢化
14. zinc [zink] *n.* 锌 (Zn)
15. nickel [ˈnikl] *n.* 镍 (Ni)
16. secondary reformer 二段（次）转化炉（器）
17. shift reaction 变换反应，转移反应
18. ethanolamine [ˌeθəˈnɔləmiːn] *n.* 乙醇胺
19. methanation [ˌmeθəˈneiʃən] *n.* 甲烷化作用
20. plump [plʌmp] *vi.* 投票赞成，坚决拥护（for）
21. one-pass 单程，非循环过程
22. opt [ɔpt] *vi.* 选择，挑选 (for, between)
23. diammonium hydrogen phosphate 磷酸氢二铵
24. cyanide [ˈsaiəˌnaid] *n.* 氰化物
25. acrylic [əˈkrilik] *a.* 聚丙烯的，丙烯酸（衍生物）的
26. nitrile [ˈnaitrail] *n.* 腈

Exercise

1. Put the following into Chinese:

   ambient　　　　hydrogenate　　azeotrope　　　desulphurisation
   bestow　　　　 methanation　　nitrogeneous　 exothermic
   embed　　　　 bench　　　　　soda ash　　　 recycle

2. Put the following into English：

   热力学　　　　锌　　　　　　磷酸氢二铵　　平衡
   自相矛盾地　　变换反应　　　蓄意地　　　　分解
   蛋白质　　　　单程　　　　　泡罩塔　　　　复分解

Reading Material

Nitric Acid and Urea

1. Nitric Acid

Production. Much of the nitrogen used by the chemical industry to make other raw materials is not used directly as ammonia, rather, the ammonia is first converted into nitric acid. Nitric acid production consumes about 20% of all the ammonia produced.

The conversion of ammonia to nitric acid is a three-stage process:

(ⅰ) $4NH_3 + 5O_2 \longrightarrow 4NO + 6H_2O$

(ⅱ) $2NO + O_2 \longrightarrow 2NO_2$

(ⅲ) $3NO_2 + H_2O \longrightarrow 2HNO_3 + NO$

The first reaction is catalyzed by platinum (in practice platinum-rhodium gauze), as can be observed on the bench with a piece of platinum wire and some concentrated ammonia solution. It might, at first sight, seem that the overall reaction to the acid would be easy; unfortunately, there are complications as nature is a good deal less tidy than chemists and engineers would prefer.

Industrially the first reaction is carried out at about 900℃ in reactors containing platinum-rhodium gauze, the temperature being maintained by the heat produced by the reaction. At these temperatures some important side reactions are also fast. Firstly, the ammonia and air mixture can be oxidized to dinitrogen and water (this reaction tends to happen on the wall of the reaction vessel if it is hot, so it needs to be deliberately cooled). Secondly, the decomposition of the first reaction product, nitric oxide, to dinitrogen and oxygen is promoted by the catalyst. It is therefore important to get the product out of the reactor as fast as possible, though this must be balanced against the need to keep the raw materials in contact with the catalyst long enough for them to react. Thirdly, the product, nitric oxide, reacts with ammonia to give dinitrogen and water, so it is important not to let too much ammonia through the catalyst beds or the result will be wasted raw material that cannot be recovered. Control of these conflicting needs is achieved by careful reactor design and by fine control of temperature and flow-rates though the reactors. The actual contact time is usually about 3×10^{-4}s.

The second and third stages have fewer complications, but both are slow and there are no known—cost-effective—catalysts. Typically, a mixture of air and nitric oxide is passed through a series of cooling condensers where partial oxidation occurs, the reaction is favored by low temperatures. The nitrogen dioxide is absorbed from the mixture as it is passed down through a large bubble-cap absorption tower; 55%~60% nitric acid emerges from the bottom.

This nitric acid cannot be concentrated much by distillation as it forms an azeotrope with water at 68% nitric acid. Nitric acid plants typically employ a tower containing 98% sulphuric acid to give 90% nitric acid from the top of the tower. Near 100% acid can be obtained if necessary by further dehydration with magnesium nitrate.

Uses of Nitric Acid. About 65% of all the nitric acid produced is reacted with ammonia to make ammonium nitrate; 80% of this is used as fertilizer, the rest as an explosive. The other major use of nitric acid is in organic nitrations. Almost all explosives are ultimately derived from nitric acid (most are nitrate esters—e.g. nitroglycerine or nitrated aromatics—e.g., trinitrotoluene). Nitration using mixtures of sulphuric and nitric acid is the first step in the synthesis of important nitro-and amino-aromatic intermediates such as aniline (the first step is nitration of an aromatic, then reduction of the nitro group to an amino). Many important dyestuffs and pharmaceuticals are ultimately derived from such reactions, though the quantities involved are small. Polyurethane plastics are built around aromatic isocyanates ultimately derived from nitrated toluene and benzene; this use consumes about 5% ~10% of nitric acid production.

2. Urea

Production. One other product of some significance is made directly from ammonia in large quantities: urea (H_2NCONH_2). About 20% of all ammonia is made into urea. It is synthesized by high pressure reaction (typically 200~400 atm and 180~210℃) of carbon dioxide with ammonia in a two-stage reaction:

(ⅰ) $CO_2 + 2NH_3 \longrightarrow NH_2CO_2^- NH_4^+$ (ammonium carbamate)

(ⅱ) $NH_2CO_2^- NH_4^+ \longrightarrow NH_2CONH_2 + H_2O$

The high pressure reaction achieves about 60% conversion of carbon dioxide to the carbamate (stage 1) and the resulting mixture is then passed into low-pressure decomposers to allow for the conversion to urea. Unreacted material is passed back to the start of the high-pressure stage of the process as this greatly improves overall plant efficiency. The solution remaining after the second stage can either be used directly as a liquid nitrogenous fertilizer or concentrated to give solid urea of 99.7% purity.

Uses. The high nitrogen content of urea makes it another useful nitrogenous fertilizer, and this accounts for the vast majority of market for the compound. Other uses are significant but use only about 10% of all the urea produced. The biggest other use is for resins (melamine-formaldehyde and urea-formaldehyde) which are used, for example, in plywood adhesives and Formica surfaces.

Selected from "An Introduction to Idustrial Chemistry, 2nd Edition, C. A. Heaton, Blackie & Son Ltd., 1997"

New Words

1. platinum ['plætinəm] *n.* 铂，白金 (Pt)
2. rhodium ['rəudiəm] *n.* 铑 (Rh)
3. gauze [gɔ:z] *n.* (金属丝，纱，线) 网
4. bench [bentʃ] *n.* 实验台，装置
5. deliberately [di'libərətli] *ad.* 故意地，蓄意地；审慎地，深思熟虑地
6. bubble-cap tower 泡罩塔
7. azeotrope [ə'zi:ətrəup] *n.* 恒沸物，共沸混合物
8. nitroglycerine [naitrəu'glisəri:n] *n.* 硝化甘油（炸药），硝酸甘油
9. trinitrotoluene [trai'naitrəu'tɔljui:n] *n.* 三硝基甲苯，TNT 炸药
10. nitration [nai'treiʃən] *n.* 硝化（作用），

渗氮（法）
11. nitro- [词头] 硝基
12. polyurethane [ˌpɔliˈjuəriθein] n. 聚氨酯，聚氨基甲酸酯
13. isocyanate [ˌaisəuˈsaiəneit] n. 异氰酸盐（酯）
14. carbamate [ˈkɑːbəmeit] n. 氨基甲酸酯
15. ammonium carbamate 氨基甲酸铵
16. nitrogenous [naiˈtrɔdʒinəs] a. 含氮的
17. melamine [ˈmeləmiːn] n. 蜜胺，三聚氰（酰）胺
18. plywood [ˈplaiwud] n. 胶合板
19. Formica [fɔːˈmaik] n. 佛米卡（一种家具表面抗热塑料贴面）

Lesson 7
Momentum, Heat, and Mass Transfer

In most of the unit operations encountered in the chemical and petroleum industries, one or more of the processes of momentum, heat, and mass transfer is involved. Thus, in the flow of a fluid under adiabatic conditions through a bed of granular particles, a pressure gradient is set up in the direction of flow and a velocity gradient develops approximately perpendicularly to the direction of motion in each fluid stream[1]; momentum transfer takes place between the fluid elements which are moving at different velocities. If there is a temperature difference between the fluid and the pipe wall or the particles, heat transfer will take place as well, and the convective component of the heat transfer will be directly affected by the flow pattern of the fluid[2]. Here, then, is an example of a process of simultaneous momentum and heat transfer in which the same fundamental mechanism is affecting both processes. Fractional distillation and gas absorption are frequently carried out in a packed column in which the gas or vapor stream rises counter-currently to a liquid[3]. The function of the packing in this case is to provide a large interfacial area between the phases and to promote turbulence within the fluids[4]. In a very turbulent fluid the rates of transfer per unit area of both momentum and mass are high; and as the pressure rises the rates of transfer of both momentum and mass increase together. In some cases, momentum, heat, and mass transfer all occur simultaneously as, for example, in a water-cooling tower, where transfer of sensible heat and evaporation both take place from the surface of the water droplets[5]. It will now be shown not only that the process of momentum, heat, and mass transfer are physically related, but also that quantitative relations between them can be developed.

Another form of interaction between the transfer processes is responsible for the phenomenon of thermal diffusion in which a component in a mixture moves under the action of a temperature gradient. Although these are important applications of thermal diffusion, the magnitude of the effect is usually small relative to that arising from concentration gradients.

When a fluid is flowing under streamline conditions over a surface, a forward component of velocity is superimposed on the random distribution of velocities of the molecules, and movement at right angles to the surface occurs solely as a result of the random motion of the molecules[6]. Thus if two adjacent layers of fluid are moving at different velocities, there will be a tendency for the faster moving layer to be retarded and the slower moving layer to be accelerated by virtue of the continuous passage of molecules in each direction. There will therefore be a net transfer of momentum from the fast to the slow moving stream[7]. Similarly, the molecular motion will tend to reduce any temperature gradient or any concentration gradient if the fluid consists of a mixture of two or more components. At the boundary the effects of the molecular transfer are balanced by the drag forces at the surface[8].

If the motion of the fluid is turbulent, the transfer of fluid by eddy motion is superimposed on the molecular transfer process. In this case, the rate of transfer to the surface will be a function of the degree of turbulence. When the fluid is highly turbulent, the rate of transfer by molecular motion will be negligible compared with that by eddy motion. For small degrees of turbulence the two may be of the same order[9].

In addition to momentum, both heat and mass can be transferred either by molecular diffusion alone or by molecular diffusion combined with eddy diffusion. Because the effects of eddy diffusion are generally far greater than those of the molecular diffusion, is occurring. Thus the main resistance to the flow of heat or mass to a surface lies within the laminar sub-layer. The thickness of the laminar sub-layer is almost inversely proportional to the Reynolds number for fully developed turbulent flow in a pipe[10]. Thus the heat and mass transfer coefficients are much higher at high Reynolds numbers.

Selected from "English for Chemical Engineers, by Ma Zhengfei etc., Southeast University Press, 2006, 63-65"

New Words

1. encounter [in'kauntə]   v. 遇见，遭遇，冲突
2. adiabatic [ædiə'bætik]   a. 绝热的，不传热的
3. granular ['grænjulə]   a. 粒状的，晶状的
4. perpendicularly [pə:pən'dikjuləli]   ad. 垂直地，正交地，直立地
5. affect [ə'fekt]   v. 影响，作用
6. promote [prəu'məut]   v. 增进，促进
7. magnitude ['mægnitju:d]   n. 巨大，重大
8. superimpose [sju:pərim'pəuz]   v. 附加，加在上面
9. retard [ri'ta:d]   v. 延缓，阻止
10. virtue ['vɜ:tʃu:]   n. 效能，效力，美德

Notes

1. A pressure gradient is set up in the direction of flow and a velocity gradient develops approximately perpendicularly to the direction of motion in each fluid stream. 参考译文：在流动方向存在压力梯度，而且在近似垂直于每股流体运动方向上形成速度梯度。

2. If there is a temperature difference between the fluid and the pipe wall or the particles, heat transfer will take place as well, and the convective component of the heat transfer will be directly affected by the flow pattern of the fluid. 参考译文：如果流体与管壁或流体与颗粒之间存在温差，那么也会出现传热，而且对流传热部分会直接受流体流型的影响。

3. The gas or vapor stream rises counter-currently to a liquid. 参考译文：气体或蒸汽相对于流体逆流上升。

4. The function of the packing in this case is to provide a large interfacial area between the phases and to promote turbulence within the fluids. 参考译文：在这种情况下，填料的作用就是在两相之间提供大的相界面积，并增强液体内部的湍动。

5. In some cases, momentum, heat, and mass transfer all occur simultaneously as, for example, in a water-cooling tower, where transfer of sensible heat and evaporation both take place from the surface of the water droplets. 参考译文：在某些情况下，动量、热量和质量传递会同时发生。例如，在凉水塔中显热（传热）和蒸发（传质）将同时在水滴表面发生。

6. Movement at right angles to the surface occurs solely as a result of the random motion of the molecules. 参考译文：垂直于壁面的运动只是由于分子的随机运动造成的。

7. There will therefore be a net transfer of momentum from the fast to the slow moving stream. 参考译文：因此将有一个净的动量从高速层传向低速层。

8. At the boundary the effects of the molecular transfer are balanced by the drag forces at the

surface. 参考译文：在边界上，分子（动量）传递的影响被壁面曳力所平衡。

9. the same order. 相同的数量级。

10. The thickness of the laminar sub-layer is almost inversely proportional to the Reynolds number for fully developed turbulent flow in a pipe. 参考译文：对于沿管道充分发展的湍流流动，层流底层的厚度几乎与雷诺数成反比。

Exercise

1. Put the following into Chinese:

encounter	magnitude	virtue	onset
conduit	adhere	finite	lateral
transition	shed light on	flask	viscous

2. Put the following into English:

绝热的	晶状的	阻止	促进
剪应力	界面张力	脉动	临界速度
层流	湍流	势流	错流

3. Comprehension and toward interpretation

   a. Do momentum, heat and mass transfer form the basis of unit operations?

   b. In some cases, momentum, heat and mass transfer all occurs simultaneously, explain with examples.

   c. What will happen for two adjacent layers of fluid with different moving velocities?

   d. Are momentum, heat and mass transfer caused by diffusion? Explain the answer.

Reading Material

Fluid-Flow Phenomena

The behavior of a flowing fluid depends strongly on whether or not the fluid is under the influence of solid boundaries. In the region where the influence of the wall is small, the shear stress may be negligible and the fluid behavior may approach that of an ideal fluid, one which is incompressible and has zero viscosity. The flow of such an ideal fluid is called potential flow and is completely described by the principles of Newtonian mechanics and conservation of mass. Potential flow has two important characteristics: (ⅰ) neither circulations nor eddies can form within the stream, so that potential flow is also called irrotational flow, and (ⅱ) friction cannot develop, so that there is no dissipation of mechanical energy into heat.

Potential flow can exist at distances not far from a solid boundary. A fundamental principle of fluid mechanics, originally stated by Prandtl in 1904, is that, except for fluids moving at low velocities or possessing high viscosities, the effect of the solid boundary on the flow is confined to a layer of the fluid immediately adjacent to the solid wall. This layer is called the boundary layer, and shear forces are confined to this part of the fluid. Outside the boundary layer, potential flow survives. Most technical flow processes are best studied by considering the fluid stream as two parts, the boundary layer and the remaining fluid. In some situations such as flow in a converging nozzle, the boundary layer may be neglected, and in others such as flow through pipes, the boundary layer fills the entire channel, and there is no potential flow.

Within the current of an incompressible fluid under the influence of solid boundaries, four important effects appear: (ⅰ) the coupling of velocity-gradient and shear-stress fields, (ⅱ) the onset of turbulence, (ⅲ) the formation and growth of boundary layers, and (ⅳ) the separation of boundary layers from contact with the solid boundary.

The Velocity Field. When a stream of fluid is flowing in bulk past a solid wall, the fluid adheres to the solid at the actual interface between solid and fluid. The adhesion is a result of the force fields at the boundary, which are also responsible for the interfacial tension between solid and fluid. If, therefore, the wall is at rest in the reference frame chosen for the solid-fluid system, the velocity of the fluid at the interface is zero. Since at distances away from the solid the velocity is finite, there must be variations in velocity from point to point in the flowing stream. Therefore, the velocity at any point is a function of the space coordinates of that point, and a velocity field exists in the space occupied by the fluid. The velocity at a given location may also vary with time. When the velocity at each location is constant, the field is invariant with time and the flow is said to be steady.

One-Dimensional Flow. Velocity is a vector, and in general the velocity at a point has three components, one for each space coordinate. In many simple situations all velocity vectors in the field are parallel or practically so, and only one velocity component, which may be taken as a scalar, is required. This situation, which obviously is much simpler than the general vector field, is called one-dimensional flow; an example is flow through straight pipe.

Laminar Flow. At low velocities fluids tend to flow without lateral mixing, and adjacent layers slide past one another like playing cards. There are neither cross-currents nor eddies. This regime is called laminar flow. At higher velocities turbulence appears and eddies form, which, as discussed later, lead to lateral mixing.

Turbulence. It has long been known that a fluid can flow through a pipe or conduit in two different ways. At low flow rates the pressure drop in the fluid increases directly with the fluid velocity; at high rates it increases much more rapidly, roughly as the square of the velocity. The distinction between the two types of flow was first demonstrated in a classic experiment by Osborne Reynolds, reported in 1883. A horizontal glass tube was immersed in a glass-walled tank filled with water. A controlled flow of water could be drawn through the tube by opening a valve. The entrance to the tube was flared, and provision was made to introduce a fine filament of colored water from the overhead flask into the stream at the tube entrance. Reynolds found that, at low flow rates, the jet of colored water flowed intact along with the mainstream and no cross mixing occurred. The behavior of the color band showed clearly that the water was flowing in parallel straight lines and that the flow was laminar. When the flow rate was increased, a velocity, called the critical velocity, was reached at which the thread of color became wavy and gradually disappeared, as the dye spread uniformly throughout the entire cross section of the stream of water. This behavior of the colored water showed that the water no longer flowed in laminar motion but moved erratically in the form of crosscurrents and eddies. This type of motion is turbulent flow.

Reynolds Number and Transition From Laminar to Turbulent Flow. Reynolds studied the conditions under which one type of flow changes into the other and found that the critical velocity, at which laminar flow changes into turbulent flow, depends on four quantities: the diameter of the tube, and the viscosity, density, and average linear velocity of the liquid. Furthermore, he found that these four factors can be combined into one group and that the change in kind of flow occurs at a definite value of the group. The grouping of variables so found was:

$$Re = \frac{D\bar{V}\rho}{\mu} = \frac{D\bar{V}}{\nu} \tag{1}$$

where D =diameter of tube
$\bar{V}$ =average velocity of liquid
μ =viscosity of liquid
ρ =density of liquid
ν =kinetic viscosity of liquid

The dimensionless group of variables defined by Eq. (1) is called the Reynolds number Re. It is one of the named dimensionless groups. Its magnitude is independent of the units used, provided the units are consistent.

Additional observations have shown that the transition from laminar to turbulent flow actually may occur over a wide range of Reynolds numbers. Laminar flow is always encountered at Reynolds numbers below 2,100 but it can persist up to Reynolds numbers of several thousand under special conditions of well-rounded tube entrance and very quiet liquid in the tank. Under ordinary conditions of flow, the flow is turbulent at Reynolds numbers above about 4,000. Between 2,100 and 4,000 a transition region is found, where the type of flow may be either laminar or turbulent, quite so and statistical analysis of the frequency distributions has proved to be useful in characterizing the turbulence.

Nature of Turbulence. Because of its importance in many branches of engineering, turbulent flow has been extensively investigated in recent years, and a large literature has accumulated on this subject. Refined methods of measurement have been used to follow in detail the actual velocity fluctuations of the eddies during turbulent flow, and the results of such measurements have shed much qualitative and quantitative light on the nature of turbulence.

Turbulent flow consists of a mass of eddies of various sizes coexisting in the flowing stream. Large eddies are continually formed. They break down into smaller eddies, which in turn evolve still smaller ones. Finally, the smallest eddies disappear. At a given time, and in a given volume, a wide spectrum of eddy sizes exists. The size of the largest eddy is comparable with the smallest dimension of the turbulent stream; the diameter of the smallest eddies is about 1 mm. Smaller eddies than this are rapidly destroyed by viscous shear. Flow within an eddy is laminar. Since even the smallest eddies contain about 10^{16} molecules, all eddies are of macroscopic size, and turbulent flow is not a molecular phenomenon.

Any given eddy possesses a definite amount of mechanical energy, much like that of a small spinning top. The energy of the largest eddies is supplied by the potential energy of the bulk flow of the fluid. From an energy standpoint turbulence is a transfer process in which large eddies, formed from the bulk flow, pass energy of rotation along a continuous series of smaller eddies. Mechanical energy is not appreciably dissipated into heat during the breakup of large eddies into smaller and smaller ones, but such energy is not available for maintaining pressure or overcoming resistance to flow and is worthless for practical purposes. This mechanical energy is finally converted to heat when the smallest eddies are obliterated by viscous action.

Selected from "Unit Operation of Chemical Engineering, 4th edition, Warren L. McCabe, McGraw-Hill, Inc., 1985"

New Words

1. shear stress   剪应力
2. negligible ['neglidʒəbl] *a.* 可忽略的，不计的，很小的
3. potential flow   势流

4. irrotational [irəu'teiʃənl]   a. 无旋的，不旋转的
5. dissipation [disi'peiʃən]   n. 耗散，损耗，消散
6. boundary layer   边界层
7. converge [kən'və:dʒ]   v. 会聚，汇合，【数】收敛
8. onset ['ɔnset]   n.（有力的）开始，发动
9. adhere [əd'hiə]   v. 黏附（于），附着（于）；坚持；追随
10. finite ['fainait]   a. 有限的，受限制的
11. tension ['tenʃən]   n. 张力，弹力
12. laminar ['læminə]   a. 层流的，层状的
13. lateral ['lætərəl]   a. 横向的，水平的
14. cross-current   n. 错流，正交流
15. conduit ['kɔndit]   n. 导管，输送管，（大）管道
16. immerse [i'mə:s]   vt. 浸（入，没），沉入；专心，埋头于，投入
17. flare [flɛə]   v. 端部张开，（向外）扩张（成喇叭形）
18. flask [fla:sk]   n. 烧瓶，长颈瓶
19. filament ['filəmənt]   n.（细）丝，（细）线；灯丝，游丝
20. mainstream ['meinstri:m]   n. 干流，主流；主要倾向
21. critical velocity   临界速度
22. erratically   ad. 不规律的，不稳定的
23. transition [træn'siʃən]   n. 过渡（段），转变，变化
24. fluctuation [flʌktju'eiʃən]   n. 脉动，波动，起伏，增减
25. shed light on   阐明，把……弄明白
26. viscous ['viskəs]   a. 黏（性，滞，稠）的
27. dissipate ['disipeit]   v. 使耗散，消除，消耗
28. obliterate [ə'blitəreit]   vt. 除去，删去，消除

Lesson 8
Heat Transfer

Practically all the operations that are carried out by the chemical engineer involve the production or absorption of energy in the form of heat. The laws governing the transfer of heat and the types of apparatus that have for their main object the control of heat flow are therefore of great importance.

1. Nature of Heat Flow

When two objects at different temperatures are brought into thermal contact, heat flows from the object at the higher temperature to that at the lower temperature. The net flow is always in the direction of the temperature decrease. The mechanisms by which the heat may flow are three: conduction, convection, and radiation.

Conduction. If a temperature gradient exists in a continuous substance, heat can flow unaccompanied by any observable motion of matter. Heat flow of this kind is called conduction. In metallic solids, thermal conduction results from the motion of unbound electrons, and there is close correspondence between thermal conductivity and electrical conductivity. In solids which are poor conductors of electricity, and in most liquids, thermal conduction results from the transport of momentum of individual molecules along the temperature gradient. In gases conduction occurs by the random motion of molecules, so that heat is "diffused" from hotter regions to colder ones. The most common example of conduction is heat flow in opaque solids, as in the brick wall of a furnace or the metal wall of a tube.

Convection. When a current of macroscopic particle of fluid crosses a specific surface, such as the boundary of a control volume, it carries with it a definite quantity of enthalpy. Such a flow of enthalpy is called a convective flow of heat or simply convection. Since convection is a macroscopic phenomenon, it can occur only when forces act on the particle or stream of fluid and maintain its motion against the forces of friction. Convection is closely associated with fluid mechanics. In fact, thermodynamically, convection is not considered as heat flow but as flux of enthalpy. The identification of convection with heat flow is a matter of convenience, because in practice it is difficult to separate convection from true conduction when both are lumped together under the name convection. Examples of convection are the transfer of enthalpy by the eddies of turbulent flow and by the current of warm air flowing across and away from an ordinary radiator.

Natural and Forced Convection. The forces used to create convection currents in fluids are of two types. If the currents are the result of buoyancy forces generated by differences in density and the differences in density are in turn caused by temperature gradients in the fluid mass, the action is called natural convection. The flow of air across a heated radiator is an example of natural convection. If the currents are set in motion by the action of a mechanical device such as a pump or agitator, the flow is independent of density gradients and is called forced convection. Heat flow to a fluid pumped through a heated pipe is an example of forced convection. The two kinds of force may be active simultaneously in the same fluid, and natural and forced convection then occur together.

Radiation. Radiation is a term given to the transfer of energy through space by electromagnetic waves. If radiation is passing through empty space, it is not transformed into heat or any other form of energy nor is it diverted from its path. If, however, matter appears in its path, the radiation will be transmitted, reflected, or absorbed. It is only the absorbed energy that appears as heat, and this transformation is quantitative. For example, fused quartz transmits practically all the radiation that strikes it: a polished opaque surface will absorb most of the radiation received by it and will transform such absorbed energy quantitatively into heat.

Monatomic and diatomic gases are transparent to thermal radiation, and it is quite common to find that heat is flowing through masses of such gases both by radiation and by conduction-convection. Examples are the loss of heat from a radiator or unlagged steam pipe to the ambient air of the room and heat transfer in furnaces and other high-temperature gas-heating equipment. The two mechanisms are mutually independent and occur in parallel, so that one type of heat flow can be controlled or varied independently of the other. Conduction-convection and radiation can be studied separately and their separate effects added together in cases where both are important. In very general terms radiation becomes important at high temperatures and is independent of the circumstances of the flow of the fluid. Conduction-convection is sensitive to flow conditions and is relatively unaffected by temperature level.

2. Rate of Heat Transfer

Heat Flux. Heat-transfer calculations are based on the area of the heating surface and are expressed in Btu per hour per square foot (or watts per square meter) of surface through which the heat flows. The rate of heat transfer per unit area is called the heat flux. In many types of heat-transfer equipment the transfer surfaces are constructed from tubes or pipe. Heat fluxes may then be based either on the inside area or the outside area of the tubes. Although the choice is arbitrary, it must be clearly stated, because the numerical magnitude of the heat fluxes will not be the same for both.

Average Temperature of Fluid Stream. When a fluid is being heated or cooled, the temperature will vary throughout the cross section of the stream. If the fluid is being heated, the temperature of the fluid is a maximum at the wall of the heating surface and decreases toward the center of the stream. If the fluid is being cooled, the temperature is a minimum at the wall and increases toward the center. Because of these temperature gradients throughout the cross section of the stream, it is necessary, for definiteness, to state what is meant by the temperature of the stream. It is agreed that it is the temperature that would be attained if the entire fluid stream flowing across the section in question were withdrawn and mixed adiabatically to a uniform temperature. The temperature so defined is called the average or mixing-up stream temperature.

3. Overall Heat-Transfer Coefficient

It is reasonable to expect the heat flux to be proportional to a driving force. In heat flow, the driving force is taken as $T_k - T_c$, where T_k is the average temperature of the hot fluid and T_c is that of the cold fluid. The quantity $T_k - T_c$ is the overall local temperature difference ΔT. It is clear that ΔT can vary considerably from point to point along the tube, and, therefore, since the heat flux is proportional to ΔT, the flux also varies with tube length. It is necessary to start with a differential equation, by focusing attention on a differential area dA through which a differential heat flow dq occurs under the driving force of a local value of ΔT. The local flux is then dq/dA and is related to the local value of ΔT by the equation

$$\frac{dq}{dA} = U\Delta T = U(T_k - T_c)$$

The quantity U, defined by equation above as a proportionality factor between dq/dA and ΔT, is called the local overall heat-transfer coefficient.

Dimensionless Groups. The values of heat-transfer coefficients for the various types of films that are encountered in convective-heat transfer are obtained from experimental data. The limited amount of data that has been obtained can be extended by means of several dimensionless groups which relate various thermal and physical properties of fluids as well as flow rates or velocities. The various dimensionless groups are combined in equations, and the experimental data are analyzed to obtain the exponents for each group and the coefficients of proportionality.

The name of a person who is being honored or who proposed the particular relationship of physical properties is usually applied to the group. Bi, Nu, Pe, Re and St are several of the more common dimensionless group used for analyzing heat transfer and for obtaining heat-transfer coefficients.

Selected from "Unit Operations of Chemical Engineering, 4th edition, Warren L. McCabe, McGraw-Hill, Inc., 1985"

New Words

1. conduction [kən'dʌkʃən] n. 传导(性,率,系数), 导热性, 导电性, 热导率, 电导率
2. unbound electron 自由电子
3. conductivity [kəndʌk'tiviti] n. 传导率, 热导率
4. diffuse [di'fju:z] v. 扩散, 散布
5. opaque [əu'peik] a. 不透明的, 不传导的, 无光泽的
6. convection [kən'vekʃən] n. (热,电)对流, 迁移
7. enthalpy [en'θælpi] n. 焓, 热焓, (单位质量的)热含量
8. flux [flʌks] n. 通量
9. eddy ['edi] n. (水、风、气等的)涡, 旋涡
10. turbulent ['tə:bjulənt] a. 湍流的, 紊流的; 扰动的
11. radiator ['reidieitə] n. 辐射体, 散热器, 暖气装置
12. buoyancy ['bɔiənsi] n. 浮力, 浮性
13. agitator ['ædʒiteitə] n. 搅拌器, 搅拌装置
14. simultaneously ad. 同时地; 同时发生地
15. divert [dai'və:t] vt. 使转向, 使变换方向, 转移
16. fuse [fju:z] v. 熔融, 熔化
17. quartz [kwɔ:z] n. 石英, 水晶
18. transparent [træns'pɛərənt] a. 透明的, 半透明的, 某种辐射线可以透过的
19. unlagged a. 未保温的, 未隔热的; 未绝缘的
20. BTU=British thermal unit 英热量单位 (=252kcal)
21. numerical [nju:'merikəl] a. 数(量、字、值)的, 用数字表示的
22. magnitude ['mægnitju:d] n. (数)量级; 大小
23. cross section (横)截面, 剖面, 断面
24. definiteness ['definitnis] n. 明确, 确定, 肯定
25. adiabatically [ædiə'bætikəli] ad. 绝热地
26. dimensionless [di'menʃənlis] a. 无量纲的
27. exponent [eks'pəunənt] n. 指数, 幂(数), 阶

Exercise

1. Put the following into Chinese:
   conductivity   agitator   opaque   regenerator

Btu　　　　　　　simultaneously　　　definiteness　　　flash evaporator
buoyancy　　　　magnitude　　　　　rigidity　　　　　recuperator

2. Put the following into English:

扩散　　　　　通量　　　　　横截面　　　　绝热地
对流　　　　　旋涡　　　　　传导　　　　　阶
焓　　　　　　湍流的　　　　无因此的　　　换热器

Reading Material

Classification of Heat Transfer Equipment

Heat transfer equipment (heat exchangers) may be defined as apparatus in which heat is transmitted from one fluid to another. The transfer of heat between the two fluids is carried out either by direct contact or by transmission through a separating wall.

According the different methods of carrying out the heat transfer operations, heat exchangers may be classified into the following three basic types:

1. Open or direct contact heat exchangers
2. Recuperators
3. Regenerators

In the first type of exchanger, the two or more phases between which heat is to be transferred are not separated by an intermediate solid wall. In the second type, heat passes through a separating wall which divides the hot fluid from the cold one. In the third type, the cold and hot fluids pass through the same space in the heat exchangers, but alternately, never at the same time.

Direct- Contact Heat Exchangers. Direct-contact heat exchangers are used when the mixing of cold and hot fluids is either harmless or desirable. They offer the potential of considerable economy due to the absence of costly heat-transfer surfaces. In order for the use of this equipment to be possible, one of two conditions must be satisfied: Either there must be a phase change on the pare of some material involved, or a second fluid must be used which normally has to be immiscible with the main process fluid.

Many types of processes utilize direct contact heat transfer. These can be divided into the following categories: 1) flash evaporators; 2) direct-contact condensers; 3) cooling tower; 4) heat transfer fluids; 5) direct heating (pneumatic drying); 6) direct use of steam; 7) submerged combustion.

Recuperator-Type Heat Exchangers. The heat exchangers used in most industrial application are of the recuperator type. In this type of equipment, heat is transferred by convection to and from the solid wall, and by conduction through the wall. The many types of recuperative heat exchangers may be classified into a number of categories, as follows:

1. By the function the heat exchangers fulfill in a process: 1) heaters, which are used primarily to heat different process fluids; 2) Coolers, which are used for cooling process fluids; 3) Condensers, whose purpose is the removal of latent rather than sensible heat.

2. By the kind of working media and their state of aggregation: 1) vapor-liquid heat exchangers; 2) liquid-liquid heat exchangers; 3) gas-liquid heat exchangers; 4) gas-gas heat exchangers.

3. By the arrangement of flow path of the working fluids: 1) unidirectional flow or parallel flow; 2) counter flow; 3) cross flow; 4) mixed flow partially (co-current and partially countercurrent).

4. By number of passes: 1) one-pass exchanger; 2) multi-pass exchanger, where the working fluids move along several passes, changing their direction of motion.

5. By material of heat transfer surface: 1) metal apparatuses, in which the heat transfer surface is made from metals; 2) non-metal apparatuses, in which the heat transfer surface is made from heat conducting non-metal materials such as ceramics or graphite.

6. By the configuration of heat transfer surfaces: 1) tubular apparatuses, in which the heat transfer surface is made of straight pipes; 2) coil apparatuses, in which the heat transfer surface is made from coils; 3) special apparatuses (plate, finned, needle-shaped, cellular, and spiral exchangers, and so on); 4) combined apparatuses, in which the heat transfer surface consists of elements of different configurations.

7. By arrangement of tubular and coiled apparatuses: 1) element apparatuses consisting of one tube or several tubes in one shell; 2) shell-and-tube exchangers, consisting of a large number of tubes enclose in one shell; 3) submerged apparatuses, consisting of elements submerged in a shell; 4) drip cooler without a shell, which is directly sprinkled by a working liquid.

8. By rigidity of construction: 1) apparatuses of a rigid structure without compensation for thermal deformation of elements; 2) apparatuses of non-rigid structure providing total compensation for thermal deformation of exchanger elements (heat exchangers with a floating head, stuffing boxes at tube plates, u-shaped tubes double-tubes and so on); 3) apparatuses with a semi-rigid structure with partial compensation for thermal deformations.

Regenerator-Type Heat Exchangers. Regenerative-type heat exchangers differ from all the previously mentioned types in that the that to be transferred is stored within the exchanger for a period of time and then removed. Either the heat-absorbing elements must remain stationary and the fluid streams must be alternately directed to it, or the elements must be moved back and forth between the passages of the hot and cold fluids. Correspondingly, regenerators may be divided into two groups on a batch or continuous basis: checker-work stationary regenerator and Lungstrom-type regenerators, where the fluid motion is continuous. Since liquids tend to adhere to metallic surfaces, a liquid film would always remain in regenerator after each cycle: therefore, regenerators are generally used only for gases. The flow of the cold fluid is usually in the opposite direction to that of the hot fluid in order to facilitate cleaning and improve efficiency.

The effectiveness of a regenerator depends upon the heat capacity of the regenerative material and its rate of heat absorption and release. Regenerators are used in blast and open-hearth furnaces, in low-temperature separation of gases by partial condensation, and as "reversing exchangers" for large-scale air-separating plants in the Fischer-Tropsch process and other synthetic processes.

Checkerwork regenerators are most commonly used for the preheating of combustion air and in open-hearth furnaces, ingot-soaking pits, glass-melting tanks, by product coke ovens, and heat-treating furnaces. The regenerators are constructed of fireclay, chrome, or silica brick shapes which are capable of absorbing and releasing considerable amounts of heat. The so-called "basket-weave" design is typical of current checkerwork designs. In blast furnaces and open-hearth regenerators, rectangular firebrick tiles are assembled in a basket-weave design to form square flues, while coke-oven tiles are made in special shape to form complicated, although often less rugged, heat-absorbing elements.

Selected from "English for Chemical Engineers, by Ma Zhengfei etc., Southeast University Press, 2006, 34-36"

Lesson 9
Distillation

Separation operations achieve their objective by the creation of two or more coexisting zones which differ in temperature, pressure, composition, and/or phase state. Each molecular species in the mixture to be separated reacts in a unique way to differing environments offered by these zones. Consequently, as the system moves toward equilibrium, each establishes a different concentration in each zone, and this results in a separation between the species.

The separation operation called distillation utilizes vapor and liquid phases at essentially the same temperature and pressure for the coexisting zones. Various kinds of device such as dumped or ordered packings and plates or trays are used to bring the two phases into intimate contact. Trays are stacked one above the other and enclosed in a cylindrical shell to form a column. Packings are also generally contained in a cylindrical shell between hold-down and support plates.

1. Continuous Distillation

The feed material, which is to be separated into fractions, is introduced at one or more points along the column shell. Because of the difference in gravity between vapor and liquid phases, liquid runs down the column, cascading from tray to tray, while vapor flows up the column, contacting liquid at each tray.

Liquid reaching the bottom of the column is partially vaporized in a heated reboiler to provide boil-up, which is sent back up the column. The remainder of the bottom liquid is withdrawn as bottoms, or bottom product. Vapor reaching the top of the column is cooled and condensed to liquid in the overhead condenser. Part of this liquid is returned to the column as reflux to provide liquid overflow. The remainder of the overhead stream is withdrawn as distillate, or overhead product.

This overall flow pattern in a distillation column provides countercurrent contacting of vapor and liquid streams on all the trays through the column. Vapor and liquid phases on a given tray approach thermal, pressure, and composition equilibriums to an extent dependent upon the efficiency of the contacting tray.

The lighter (lower-boiling) components tend to concentrate in the vapor phase, while the heavier (higher-boiling) components tend toward the liquid phase. The result is a vapor phase that becomes richer in light components as it passes up the column and a liquid phase that becomes richer in heavy components as it cascades downward. The overall separation achieved between the distillate and the bottoms depends primarily on the relative volatilities of the components, the number of contacting trays, and the ratio of the liquid-phase flow rate to the vapor-phase flow rate.

If the feed is introduced at one point along the column shell, the column is divided into an upper section, which is often called the rectifying section, and a lower section, which is often referred to as the stripping section. These terms become rather indefinite in multiple-feed columns and columns from which a product sidestream is withdrawn somewhere along the column length in addition to the two end-product streams.

Equilibrium-Stage Concept. Energy and mass-transfer processes in an actual distillation column are much too complicated to be readily modeled in any direct way. This difficulty is circumvented by the equilibrium-stage model, in which vapor and liquid streams leaving an equilibrium stage are in complete equilibrium with each other and thermodynamic relations can be used to determine the temperature and relate the concentrations in the equilibrium streams at a given pressure. A hypothetical column composed of equilibrium stages (instead of actual contact trays) is designed to accomplish the separation specified for the actual column. The number of hypothetical equilibrium stages required is then converted to a number of actual trays by means of tray efficiencies, which describe the extent to which the performance of an actual contact tray duplicates the performance of an equilibrium stage.

Use of the equilibrium-stage concept separates the design of a distillation column into three major steps: (i) Thermodynamic data and methods needed to predict equilibrium-phase compositions are assembled. (ii) The number of equilibrium stages required to accomplish a specified separation, or the separation that will be accomplished in a given number of equilibrium stages, is calculated. (iii) The number of equilibrium stages is converted to an equivalent number of actual contact trays or height of packing, and the column diameter is determined.

All separation operations require energy input in the form of heat or work. In the conventional distillation operation, energy required to separate the species is added in the form of heat to the reboiler at the bottom of the column, where the temperature is highest. Also, heat is removed from a condenser at the top of the column, where the temperature is lowest. This frequently results in a large energy-input requirement and low overall thermodynamic efficiency. With recent dramatic increases in energy costs, complex distillation operations that offer higher thermodynamic efficiency and lower energy-input requirements are being explored.

Related Separation Operations. The simple and complex distillation operations just described all have two things in common: (i) both rectifying and stripping sections are provided so that a separation can be achieved between two components that are adjacent in volatility; and (ii) the separation is effected only by the addition and removal of energy and not by the addition of any mass separating agent (MSA) such as in liquid-liquid extraction. Sometimes, alternative single- or multiple-stage vapor-liquid separation operations may be more suitable than distillation for the specified task.

2. Batch Distillation

A batch still showing the separation of A and B.

Heating an ideal mixture of two volatile substances A and B (with A having the higher volatility, or lower boiling point) in a batch distillation setup (such as in an apparatus depicted in the opening figure) until the mixture is boiling results in a vapor above the liquid which contains a mixture of A and B. The ratio between A and B in the vapor will be different from the ratio in the liquid: the ratio in the liquid will be determined by how the original mixture was prepared, while the ratio in the vapor will be enriched in the more volatile compound, A (due to Raoult's Law, see above). The vapor goes through the condenser and is removed from the system. This in turn means that the ratio of compounds in the remaining liquid is now different from the initial ratio (i.e. more enriched in B than the starting liquid).

The result is that the ratio in the liquid mixture is changing, becoming richer in component B. This causes the boiling point of the mixture to rise, which in turn results in a rise in the temperature in the vapor, which results in a changing ratio of A:B in the gas phase (as distillation continues, there is an increasing proportion of B in the gas phase). This results in a slowly changing ratio A:B in the distillate.

If the difference in vapor pressure between the two components A and B is large (generally expressed as the difference in boiling points), the mixture in the beginning of the distillation is highly enriched in component A, and when component A has distilled off, the boiling liquid is enriched in component B.

Selected from "Chemical Process Equipment, Stanley M. Walas, Butterworth Publlishers,1988"

New Words

1. dumped packing   乱堆填料
2. ordered packing   整砌填料，规整填料
3. hold-down   *n.* 压具（板，块），压紧（装置），固定
4. feed [fi:d]   *n.* 进料，加料；加工原料
5. cascade [kæsˈkeid]   *v.; n.* 梯流，阶流式布置；级联，串联
6. boil-up   蒸出（蒸汽）
7. bottom [ˈbɔtəm]   *n.(pl.)* 底部沉积物，残留物，残渣
8. relative volatility   相对挥发度（性）
9. rectify [ˈrektifai]   *vt.* 精馏，精炼，蒸馏
10. rectifying section   精馏段
11. stripping [ˈstripiŋ]   *n.* 洗提，气提，解吸
12. stripping section   提馏段
13. multiple-feed   多口进料
14. sidestream   *n.* 侧线馏分，塔侧抽出物
15. circumvent [ˌseːkəmˈvent]   *vt.* 绕过，回避，胜过
16. hypothetical [ˌhaipəˈθetikl]   *a.* 假定(设，说) 的，有前提的
17. duplicate [ˈdjuːplikeit]   *vt.* 重复，加倍，复制
18. mass separating agent   质量分离剂

Exercise

1. Put the following into Chinese:
   dumped packing   ordered packing   sidestream   mass separating agent
   coexisting   stripping   boil-up   insoluble
   cascade   rectify   boiling point   entrainer
2. Put the following into English:
   相对挥发度   精馏段   提馏段   压具
   二元的   进料   假设   支撑板
   残留物   复制   恒沸精馏   萃取精馏

Reading Material

Azeotropic and Extractive Distillation

Numerous processes have been developed which separate liquids by distillation with some modifications, or using an additional physical or chemical mechanism.

Azeotropic Distillation. In its simplest form, azeotropic distillation depends on the use of an added liquid—the entrainer. This is usually infinitely miscible with one of the components to be separated. The entrainer is quite insoluble in the other component, with which it forms an azeotropic mixture. This has a total pressure equal to the sum of the partial pressures of the two, and a boiling point—the azeotropic boiling point—which is much lower than that of either of the original liquids of the mixture. Thus the entrainer brings over this azeotropic mixture as the vapors at the column head in what is essentially a steam distillation. These vapors condense in the condenser, the condensate separates into two layers because of the mutual insolubility, the insoluble entrainer layer is returned to the columns as reflux, and the layer of the one component of the original mixture is removed as product.

Azeotropic distillation was developed by Young at the turn of this century using benzene as the entrainer to separate the last of the water from its hydroxyl-homologue, ethanol, which has very similar properties. The literature since is full of related data for the azeotropes of other liquids with alcohol which were found in attempts to find something other than benzene. All of these, however, like benzene, increase the volatility or lower the boiling point of the ethanol as well as that of the water. Many people searched for one which would lower the boiling point only of the water. Wentworth did indeed find the liquid to simplify the process of removal of the last of the water by lowering the effective boiling point of only the water. This system is now available in a considerably improved form, which will give at the lowest cost the large amounts of absolute alcohol being planned for use motor fuel in this country as well as in other countries of the world.

Azeotropic distillation was developed industrially for separating water from another hydroxy-relative, acetic acid, in another series, again by effectively making the water more volatile. This was before the days of stainless steel. Copper and bronze were then the accepted materials of construction, with special parts and condensers being made of silver, often many thousands of ounces in a single unit. In a few years, to accomplish this separation at one plant, probably the largest mass of copper ever to be used in one area was in huge distillation units of massive construction to withstand the inevitable corrosion. Numerous other plants in this and other countries use this process for the separation of water from the spent acetic acid from synthesis processing where acetic may be not only an acid in the reaction but also a solvent for reactants or products.

Extractive Distillation. As just mentioned, a selected third liquid is added to a mixture of two other liquids for an azeotropic distillation to increase the vapor pressure or effectively to lower the boiling point of the one of these two in which it is largely insoluble. Contrariwise, for an extractive distillation a selected third liquid, of a higher boiling point, is added to the top of a column in which is being separated a mixture of two other liquids. With one of these, the added liquid has large mutual solubility, i.e., for this one it acts as a solvent. The high boiling solvent carries the liquid down and out the bottom, while the other liquid goes, almost, pure in the vapors off the top.

Over 50 years ago this method was used in several installations in central Europe, two in the United States, to separate that difficult pair of liquids — acetic acid and water. It has also been used more recently in separating hydrocarbons by class, i.e., aromatics from paraffins, rather than as in ordinary distillation by volatilities. Also, liquid mixtures, as acetone and methanol, which cannot otherwise be separated by distillation may be separated by extractive distillation, as may some of the impurities obtained in fermentation of carbonhydrates to alcohol.

Another variety of extractive distillation involves the use of a salt in solution as the solvent. This effectively reduces the vapor pressure of one component of a mixture — such as water or a hydroxyl relative, i.e., a lower alcohol. Professor W. F. Furter has been a protagonist for such separations both as to the determination of the underlying physical data with his coworkers, and then their use in designing separation processes.

Selected from "English for Chemical Engineers, by Ma Zhengfei etc., Southeast University Press, 2006, 45-46"

Lesson 10
Gas Absorption

Gas absorption is an important unit operation in chemical engineering, particularly in the heavy chemical industry. In the purification of coal gas and synthetic gas as well as in the manufacture of hydrochloric acid, sulphuric acid, nitric acid, soda ash and bleaching lyes, to mention only the most striking examples, gas absorption is the essential feature of the process. Since the twenties extensive research has been carried out with the result that the knowledge of this unit operation has increased considerably.

The essential element in gas absorption is the mass transfer between two fluid phases, if necessary, combined with chemical reactions. The latter are mainly restricted to the liquid phase although, for instance in the manufacture of nitric acid, the chemical reaction may also occur in the gas phase.

It will be understood that in such a heterogeneous process as gas absorption a large contact area is a primary requisite for rapid transfer.

There are three ways in which a large contact area can be established:
1. The liquid is brought in contact with the gas in the form of thin films (film scrubbers).
2. The liquid is dispersed in the gas in the form of minute drops (spray scrubbers).
3. The gas is dispersed in the liquid in the form of small bubbles (bubble scrubbers).

All apparatus applied in gas absorption practice is based on one of these three principles or on a combination of them.

Here serious problems are encountered which have only been partly solved. The first things to discover when dealing with a process of mass transfer are the size of the contact area and the way in which this size depends on the various condition of the experiment. The most complete answer to this question can be even for apparatus of type 1. The simplest representative of this type is the wetted wall column, the contact area of which may be determined with great accuracy. For packed column the question cannot be answered so easily, since it cannot be exactly ascertained which part of the packing will be wetted. For the sake of simplicity the contact area may be considered to be equal to the total area of the packing material. In apparatus in which gases or liquids are dispersed, type 2 and 3, determination of the contact area is almost impossible.

Only in the last few years has insight been obtained into the conduct of gas bubbles during their formation in an ascension through liquids, although much work is still to be done in this field.

Absorption, or gas absorption, is a unit operation used in the chemical industry to separate gases by washing or scrubbing a gas mixture with a suitable liquid. One or more of the constituents of the gas mixture will dissolve or be absorbed in the liquid and can thus be removed from the mixture. In some systems, this gaseous constituent forms a physical solution with the liquid or the solvent, and in other cases, it reacts with the liquid chemically.

The purpose of such scrubbing operation may be any of the following: gas purification (e.g., removal of air pollutants from exhaust gases or contaminants from gases that will be further processed), product recovery and production of solutions gases for various purposes.

Gas absorption is usually carried out in vertical countercurrent columns. The solvent is fed at the top of the absorber, whereas the gas mixture enters from the bottom. The absorbed substance is washed out by the solvent and leaves the absorber at the bottom as a liquid solution. The solvent is often recovered in a subsequent stripping or desorption operation, which is essentially the reverse of absorption. The absorber may be a packed, plate, or simple spray column. The packed column is a shell filled with packing material designed to disperse the liquid and bring it into close contact with the rising gas. In plate towers, liquid flows from plate to plate in cascade fashion and gases bubble through the flowing liquid at each plate through a multitude of dispersers or through a cascade of liquid.

The advantages of packed columns include simpler and cheaper construction. They are preferred for corrosive gases because packing, but not plates, can be made, e.g., from ceramics. They are also used in vacuum applications because the pressure drop is usually less than through plate columns.

Plate absorbers are used in applications where tall columns are required, because tall packed tower are subject to channeling and maldistribution of the liquid streams. Plate tower can be more easily cleaned. They are also preferred in applications with large heat effects requiring internal cooling since coils are more easily installed in plate towers; they can also be designed for large liquid holdup.

The fundamental physical principles underlying the process of gas absorption are the solubility of the absorbed gas and the rate of mass transfer. Information on both must be available when sizing the equipment for a given application.

In addition to the fundamental design concept (solubility and mass transfer), many practical details have to be considered during actual design and construction which may affect the performance of the absorber significantly.

Selected from "Specialized Chemical English (unformal published) (the first volumn), by Cui Bo etc., 1994, 121-128"

New Words

1. scrub [skrʌb] v. 使（气体）净化；洗气；洗涤
2. solvent ['sɔlvənt] n. 溶剂, a. 有溶解力的
3. pollutant [pə'luːtənt] n. 污染物
4. exhaust [igˈzɔːst] n. 排出, 排气; vt. 用尽, 排出
   exhaust gas 废气
5. vertical [ˈvɜːtikl] a. 垂直的
6. countercurrent [ˈkauntəˈkʌrənt] a. 逆流的, 对流的
7. recover [riˈkʌvə] vt. 回收, 复原
8. absorber [əbˈsɔːbə] n. 吸收器
9. strip [strip] vt. 解吸, 汽提
10. desorption [diːˈsɔːpʃən] n. 解吸作用
11. shell [ʃel] n. 壳
12. fill [fil] vt. 填充, 装满
13. cascade [kæsˈkeid] n. 阶式
14. fashion [ˈfæʃən] n. 型, 样式
15. multitude [mʌltitjuːd] v.; n. 众多, 大量
16. disperser [disˈpəːsə] n. 分散器, 泡罩
17. prefer [priˈfəː] n. 宁愿（选择）, 更喜欢
18. ceramics [siˈræmiks] n. 陶瓷制品, 陶瓷学
19. channel [ˈtʃænl] v. 沟流; n. 沟流槽
20. maldistribution [ˌmældistriˈbjuːʃən] v.; n. 分布不均, 分布不当
21. install [inˈstɔːl] vt. 安装, 设置
22. holdup [ˈhəuldʌp] n. 容纳量, 塔储量
23. underlie [ˌʌndəˈlai] vt. 作为……的基础

underlay [ˌʌndəˈlei] vt. 作为……的基础
underlain [ˌʌndəˈlein] vt. 作为……的基础
underlying [ˌʌndəˈlaiiŋ] a. 作为……的基础
24. solubility [sɔljuˈbiliti] n. 溶(解)度，可溶性
25. concept [ˈkɔnsept] n. 概念，观念，思想
26. absorption [əbˈzɔːpʃən] n. 吸收，吸收作用
27. synthetic [sinˈθetik] a. 合成的
28. hydrochloric [ˌhaidrəˈklɔrik] a. 氯化氢的
    hydrochloric acid 盐酸
29. sulphuric [sʌlˈfjuərik] a. 硫的
    sulphuric acid 硫酸
30. soda [ˈsəudə] n. (钠)碱，苏打，碳酸钠
31. ash [æʃ] n. 灰，粉
    soda ash 纯碱，苏打灰，碳酸钠
32. bleach [ˈbliːtʃ] v. 漂白
33. lye [lai] n. 碱液
34. striking [sˈtraikiŋ] a. 引人注目的，显著的
35. feature [ˈfiːtʃə] n. 要点，特征
36. extensive [iksˈtensiv] a. 广泛的
37. mass [ˈmæs] n. 质量
38. phase [feis] n. 相，状态
39. restrict [risˈtrikt] vt. 限制
40. heterogeneous [ˌhetərəˈdʒiːniəs] a. 非均相的，不均的
41. primary [ˈpraiməri] a. 主要的，基本的

（多相的）
42. requisite [ˈrekwizit] n. 必要的
43. establish [isˈtæbliʃ] vt. 建立
44. scrubber [ˈskrʌbə] n. 洗涤器
45. disperse [disˈpəːs] v.（使）分散，扩散
46. minute [ˈminit] a. 微小的
47. spray [ˈsprei] n. 喷雾，喷淋 v. 喷
48. bubble [ˈbʌbl] n. 气泡 v.（使）起泡
49. apparatus [æpəˈreitəs] n. 设备
50. encounter [inˈkauntə] vt. 遇到
51. representative [repriˈzentətiv] n. 代表，典型
52. wet [wet] vt. 把……弄湿 a. 湿的
53. accuracy [ˈækjurəsi] n. 精确，准确
54. pack [pæk] v. 填料，填塞，装重
55. ascertain [æsəˈtein] vt. 弄清，确定
56. sake [seik] n. 缘故
57. simplicity [simˈplisiti] n. 简单，简易
58. packing [ˈpækiŋ] n. 填料，填充物
59. determination [ditəːmiˈneiʃən] n. 测定，确定
60. insight [ˈinsait] n. 洞悉，洞察
61. conduct [ˈkɔndəkt] n. 行为
62. be subject to 易受
63. ascension [əˈsenʃən] n. 上升，升高

Exercise

1. Put the following into Chinese:
   solvent   weir   packed tower   representative
   vertical   spary column   bubble-plate   absorption
   recover   counter current primary   be subject to
2. Put the following into English:
   污染物   回流   非均相的   广泛的
   废气   溶解度   解吸   沟流
   逆流的   纯碱   相态   填料

Reading Material

Principle Types of Absorption Equipment

Industrial apparatus for gas absorption may usually be classified as one of four quite different types, each having as a principle objective the promotion of interphase contact between gas and liquid. Many varieties and combinations of these types exits or are possible, but only the major classifications will be described briefly.

Spray Towers consist of large empty chambers through which the gas circulates and into which the liquid is introduced in the form of droplets by means spray nozzles or other atomizing devices. The sprays may be introduced at the top of a cylindrical tower and the gas passed in at the bottom. The contact is not truly countercurrent, however, as the momentum of the injected spray is usually sufficient to stir the gas thoroughly in a tower only a few diameters high. As a result, the composition of the gas is nearly uniform throughout the chamber. The spray nozzles break the liquid into a large number of small drops, providing the interfacial surface across which diffusional transfer takes place. Within the smallest drops the liquid is stationary, and movement of the solute, takes place by molecular diffusion. The larger drops of the spray may be mixed internally however, on account of liquid circulation caused by frictional drag at the drop surface.

Although diffusion is slow inside the smaller drops, the continuous formation of fresh liquid surface at the spray nozzles allows absorption to take place rapidly. The spray is formed by the collapse of high-velocity liquid jets at the nozzles, and the gas-film resistance around the drops is relatively small because of the high velocities with which the drops are propelled into the gas. The interfacial area present in a spray chamber is surprisingly small, so that the rate of absorption per unit volume of chamber may be smaller than for other types of equipment.

Spray towers is most useful for the absorption of relatively insoluble gases, where the liquid-phase resistance controls the rate of mass transfer. The aeration of sewage is a typical application of a porous-plate device. Other types of equipment may be more economical when expensive power must be supplied to overcome the large liquid head in introducing the gas below the surface.

Bubble-Plate and Sieve-Plate Absorbers are commonly used in industry. They represent a case intermediate between simple spray chambers and aerated-tank absorbers. In them, bubbles are formed at the bottom of a shallow pool of liquid by forcing the gas either through a metal plate drilled or punched with many small holes or under a number of slotted, bell-shaped metal caps immersed in the liquid. A large share of the interphase transfer occurs as the gas bubbles are formed and as they rise through the agitated liquid. Additional transfer takes place above the liquid surface, owing to the spray and foam which are thrown up by the violent mixing of liquid and vapor on the plate. Such plates or trays are arranged one above another in a cylindrical shell. The liquid flows downward, crossing first one plate and then the next below. The vapor rises through the plates. This type of equipment has the practical advantage that the gas stream is dispersed nearly uniformly through all the liquid. There are no stagnant zones which are by-passed by one of the fluid streams, and channeling is avoided. Plate-type absorbers are used frequently in the petroleum and chemical industries, especially for absorbing soluble gases.

Packed Tower is the fourth general type of equipment, in which the liquid stream is subdivided to provide a large interfacial area as it flows by gravity over the surface of a packing material. A large number of different types of "packing" are in use, ranging from crushed stone through specially fabricated hollow ceramic cylinders to wood or ceramics salts formed into grinds. The liquid flows down the packing surface in thin films or individual streams, without filling the void space within the packing. The gas may flow downward, parallel to the liquid, or upward. Both the liquid and the gas phases are well agitated, and the equipment of this type may be used for absorbing either soluble or relatively insoluble gases. Some difficulty is experienced in maintaining uniform liquid and gas flow throughout a cross section of a packed tower, so that the operation may not be truly countercurrent. Nevertheless, equipment of this type is used frequently, especially where low pressure drop is required or where corrosive fluids are encountered.

Selected from "English for Chemical Engineers, by Ma Zhengfei etc., Southeast University Press, 2006, 40-41"

Lesson 11
Liquid Extraction

Process for the separation of the components of a solution which depend upon the unequal distribution of the components between two immiscible liquids are known as liquid-liquid extraction, or more simply as liquid extraction, processes, and sometimes as solvent extraction (although the last is frequently also applied to the process of leaching a soluble substance from a solid with a liquid solvent).

The process may be carried out in a number of ways. In most instances, the liquid solution to be treated is contacted intimately with a suitable incompletely miscible liquid which preferentially extracts one or more components. For example, acetone may be preferentially extracted from solution in water by contacting the water with chloroform. The chloroform then is found to contain a large part of the acetone, but little water. Sometimes two immiscible liquids are used, for example, a solution of acetone and acetic acid may be separated by distributing the components between water and chloroform. In this case, the acetone preferentially enters the chloroform while the acetic acid preferentially enters the water. Separations of this sort are essentially physical in character, and the various components are unchanged chemically. Nevertheless, the chemical nature of the liquids strongly influences the extent of separation possible, since the distribution of a solute depends on the extent of non-ideality of the solutions involved. Furthermore, the non-ideality may be altered in order to influence the distribution favorably, for example, by changing the temperature, by addition of salt to "salt out" an organic solute from a water solution, or by adjusting the pH of the solution or the state of oxidation of a metallic solute. In other cases, chemical reaction in the extracting liquid is actually resorted to, as in the extraction of phenol from light oil by aqueous caustic solutions.

Fields of Usefulness. Because liquid extraction results in a new solution which in turn must be separated, the more direct method of separation by distillation is usually considered first. However, liquid extraction separated primarily according to chemical type, and is therefore capable of separations which are impossible by ordinary distillation means. Thus, aromatic and paraffinic hydrocarbons of the same boiling range may be separated by liquid extraction with diethylene glycol or sulfolane as extractants for the aromatics. Even solutions which can be separated by distillation, but which are expensive to deal with in this manner, may frequently be separated better by liquid extraction. For example, distillation of a dilute solution of acetic acid in water involves the vaporization of large amounts of water at high reflux ratio, and is expensive because of the high latent heat of vaporization of water. Extraction of the acid into ethyl acetate, followed by distillation on the new solution thus formed, provides a cheaper process. Similarly, costly evaporation of water from a non-volatile solute may sometimes be circumvented by extraction of the solute into a solvent of small latent heat. Costly fractional crystallization may be avoided, as in the separation of tantalum and columbium in solution by liquid extraction, which is relatively easy. In this case, even chemical methods of separation are impractical. Extensive application in the field of process metallurgy has been made, as in the separation of uranium-vanadium, hafnium-zirconium, recovery of copper from dilute solution, and many other examples. Heat-sensitive substances, such as penicillin, may be separated from the mixtures in which they are formed by extraction into a suitable solvent at low temperature. While liquid extraction involves transfer matter between two immiscible liquids, the transfer of heat may be done by the same techniques, and direct contact of two immiscible liquids for this purpose provides a non-fouling heat exchanger of increasing interest in recent years.

Definitions. The solution whose components are to be separated is the feed to the process. Liquid added to the feed for purpose of extraction is the solvent. If the solvent consists primary of one substance (a side from small amounts of residual feed material which may be present in a recycled, recovered solvent), it is a single solvent. A solvent consisting of a solution of one or more substances chosen to provide special properties is a mixed solvent. The solvent lean, residual feed solution, with one or more constituents removed by extraction, is the raffinate. The solvent-rich solution containing the extracted solute(s) is the extract. Two immiscible solvent between which the feed constituents distribute is a double solvents in which case the terms extract and raffinate no longer apply.

The minimum requirement for liquid extraction is the intimate contact of two immiscible liquids for the purpose of mass transfer of constituents from one liquid (or phase) to the other, followed by physical separation of the two immiscible liquids. Any device or combination of devices which accomplishes this once is a stage. If the effluent liquids are in equilibrium, so that no further change in concentration would have occurred within them after longer contact time, the stage is a theoretical or ideal stage. The approach to equilibrium actually attained is the stage efficiency. A multistage cascade is a group of stages, arranged for countercurrent or other type of flow of liquids from one to the other for purposes of enhancing the extent of separation.

In countercurrent flow particularly, the liquids are frequently contacted without repeated physical separation and recontacting in discrete stages. Such methods are known as differential-or continuous-contacting methods.

Solvent Recovery. It is usually desired that the extracted solutes be ultimately obtained free of solvent. In any case, cost considerations almost always require that the solvent be recovered for reuse, not only from the extract but in most cases from the raffinate as well even though the solvent content of the latter may be relatively small. Solvent recovery is usually accomplished by distillation, as in Fig. 11-1(a), which is typified by the recovery of acetic acid from a dilute aqueous feed with ethyl acetate as solvent. The distillation operations are shown only schematically, and will vary considerably in detail depending upon whether the solvent is high or low-boiling, whether azeotropes (ordinary or heterogenous two liquid phase) form with the solvent, and relative solubility and volatility of the solvent. The details are discussed by Treybal ("Liquid Extraction," 2d ed., McGraw-Hill, New York, 1963). Sometimes the extracted solute is desired in solution form. Figure 11-1(b) schematically represents such a process, using liquid extraction for solvent recovery, typified by the recovery of uranium from an ore leach-liquor feed. Here the feed, contacting uranium together with undesired metals, is extracted with a uranium-complexing solvent such as tributyl phosphate dissolved in kerosene, under such conditions of pH and state of oxidation that only uraniumid extracted. The raffinate contains so little solvent that it needs no treatment for solvent recovery. The extract is contacted in the solvent stripper (by liquid extraction methods) with aqueous acid, whereupon the uranium enters the acid and the solvent is restored to its original condition for reuse. In this way, the uranium is concentrated several-hundred-fold, as well as separated from impurities.

Selected from "Specialized Chemical English (unformal published) (the first volumn), by Cui Bo etc., 1994, 173-179"

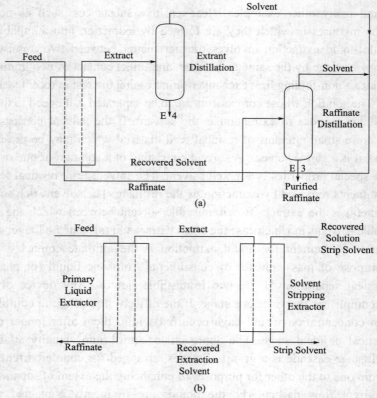

Fig.11-1 Schematic flow sheets of liquid-extraction processes with associated solvent recovery by (a) distillation and (b) liquid extraction

New Words

1. extraction [ik'strækʃn] n. 萃取
2. immiscible [i'misibl] a. 不（能）混合的，非互溶的
3. intimately ['intimitli] ad. 密切的，紧密的
4. acetone ['æsitəun] n. 丙酮
5. chloroform ['klɔrəfɔ:m] n. 三氯甲烷，氯仿
6. acetic acid n. 醋（乙）酸
7. phenol ['fi:nɔl] n. 酚醛树脂
8. tantalum ['tæntələm] n. 钽
9. columbium [kə'lʌmbiəm] n. 铌
10. uranium [juə'reiniəm] n. 铀
11. vanadium [və'neidjəm] n. 钒
12. hafnium ['hæfniəm] n. 铪
13. zirconium [zə'kəunjəm] n. 锆
14. raffinate ['ræfineit] n. 提余液，残液

Exercise

1. Put the following into Chinese:
   immiscible   acetone   non-fouling   impractical
   intimately   aqueous   extract   purification
   preferentially   caustic   effluent   volatility

2. Put the following into English:
   萃取   苯酚   残液   汽提塔
   浸取   盐析   共沸物   热敏性
   乙酸   回流比   串联   青霉素

Reading Material

The Industrial Application of Liquid-Liquid Extraction

The separation of the constituents of a homogeneous liquid mixture is a problem frequently encountered in the chemical processing industry. Liquid-liquid extraction has been emerging as a very important method for separating such liquid mixtures.

The case for the use of liquid-liquid extraction will depend upon its either accomplishing a separation that cannot be achieved by other operations such as distillation, evaporation and crystallization, or effecting the separation more economically. Liquid-liquid extraction is now being adopted as a more economic alternative to other separation processes and has found immense applications in the separation of:

1. solutions of components having low relative volatility, especially when vacuum distillation is expensive,
2. solutions of close boiling and azeotrope-forming components,
3. dissolved solute when evaporation may be impractical,
4. solutions of heat-sensitive components such as antibiotics,
5. components of differing chemical type whose boiling points may overlap as in the case of petroleum hydrocarbons.

The chemical industries in which liquid-liquid extraction processes are adopted may be broadly classified under organic and inorganic chemical industries.

Organic Chemical Industries. The liquid-liquid extraction technique has been used for separating and purifying organic compounds of commercial importance on a large scale. The first recorded application of solvent extraction is probably that of Goering in 1883 for concentrating acetic acid from dilute solutions by extraction with a solvent such as ethyl acetate. However, it was in the petroleum industry that one of the first large-scale applications of liquid extraction was made, viz., the separation of aromatic compounds from aliphatic compounds.

In 1908 Edeleanu used liquid sulphur dioxide to extract the smoke-forming aromatics from Rumanian illuminating oil. The Edeleanu process has since then gained importance. As an alternative to this process the UDEX process originally developed by Dow Chemical Company has gained prominence. In this process a mixture of glycols and water is used as solvent for the separation of high purity aromatics from the mixtures of other hydrocarbons.

Another widely used application of solvent extraction in the petroleum industry is in dewaxing and deasphalting lubricating oils using liquid propane. The deasphalted oil extract obtained is chilled for crystallizing out the wax which remains suspended in propane-oil solution. The wax is then separated by filtration.

Liquid extraction has also been used extensively in the coal tar industry for many years. Extraction of tar acid from crude tar distillates has been practised commercially for a long time using aqueous solution of sodium hydroxide. High-purity tar acids are successfully recovered by fractional extraction using a double solvent, aqueous methanol and hexane. Just as in the petroleum industry, the UDEX process using aqueous diethylene glycol as solvent is adopted for extracting high purity aromatics from coke-oven oil, phenols from coke-oven gas liquors are separated using benzene as solvent.

Liquid extraction has become an important method in the refining of vegetable oils and animal fats. In the SOLEXOL process propane is used as the solvent and this process makes use of the property of varying solubilities of different constituents of oils and fats in the solvent at different temperatures.

The separation and purification of pharmaceutical and natural products have also been made possible by using liquid-liquid extraction technique. This technique is extensively used in the recovery of antibiotics from fermentation broths as most of products are heat sensitive.

With the industrially developed countries becoming more and more conscious of industrial effluent pollution of rivers and seas, solvent extraction is being used advantageously for the recovery of polluting chemicals. Phenols from catalytic petroleum cracking plant effluent streams are recovered using light catalytic oil by a process know as PHENEX process. Valuable acetic and formic acids are recovered from pulp mill black liquor using methyl-ethyl ketone.

Inorganic Industries. Liquid-liquid extraction processes have been increasingly used in the metallurgical industries. Research carried out on the processing of nuclear fuels has yielded a vast amount of information on the extraction of many metals. The rare metal salt is usually present in the leach liquor as a halide, sulphate or nitrate with an excess of the corresponding acid. Typical examples of solvents used for extracting rare metal salts are diethyl-ether, mono-, di-and tri-butyl phosphates, di-octyl phosphate, methyl ethyl ketone and higher alkyl amines.

Several processes have been developed to separate uranium from gangue metals in the uranium ore-leach liquor and also to recover unused fuel from the spent nuclear fuel.

Thorium is produced from monazite sands using solvent extraction procedure similar to that of uranium. Monazite may be reacted with sulphuric acid and the thorium precipitated as oxalate and dissolved in nitric acid. The thorium nitrate in nitric acid solution is then extracted with tributyl phosphate in xylene or kerosene. Di-tridecyl amine can also be used for the extraction of thorium from dilute solutions of thorium nitrate.

Reactive grade zirconium has been produced by removing its chemical homologue hafnium by solvent extraction process using a mixed solvent hexone and tri-butyl phosphate.

The result of research employing liquid-liquid extraction in nuclear industry are now being extended to other metallurgical industries especially in the field of hydrometallurgy. Most of these applications depend upon dissolving some organic compound in an organic solvent to form a complex with one or more metal ions whereby the distribution coefficients of the metals in the solvent are altered.

A more interesting application of solvent extraction is in the fertilizer industry. The normal process for producing phosphoric acid from the rock phosphate needs sulphuric acid but IMI (Israel Mining Industries) process uses hydrochloric acid. The digested liquor thus will contain both phosphoric acid and calcium chloride phosphoric acid is then extracted from the reaction mixture using a $C_4 \sim C_5$ industrial alcohol. By controlling the pH, the phosphoric acid can be subsequently back-washed into aqueous solution. The phosphoric acid obtained by this method is found to be very pure.

The industrial applications indicated above show the importance of liquid-liquid extraction and its great potential as a separation and purification method in chemical engineering operations.

Selected from "English for Chemical Engineers, by Ma Zhengfei etc.,Southeast University Press, 2006, 50-52 "

Lesson 12

Surface Chemistry and Adsorption

Even the most carefully polished surfaces are not smooth in a microscopic sense, but are irregular, with valleys and peaks alternating over the area. The regions of irregularity are particularly susceptible to residual force fields. At these locations the surface atoms of the solid may attract other atoms or molecules in the surrounding gas or liquid phase. Similarly, the surfaces of pure crystals have nonuniform force fields because of the atomic structure in the crystal. Such surfaces also have-sites or active centers where adsorption is enhanced. Two types of adsorption may occur. Physical adsorption is nonspecific and somewhat similar to the process of condensation. The forces attraction the fluid molecules to the solid surface are relatively weak, and the heat evolved during the exothermic adsorption process is of the same order of magnitude as the heat of condensation, 0.5 to 5 kcal/g mol. Equilibrium between the solid surface and the gas molecules is usually rapidly attained and easily reversible, because the energy requirements are small. The energy of activation for physical adsorption is usually no more than 1 kcal/g mol, since the forces involved in physical adsorption are weak. Physical adsorption cannot explain the catalytic activity of solids for reactions between relatively stable molecules, because there is no possibility of large reductions in activation energy. Reactions of atoms and free radicals at surfaces sometimes involve small activation energies, and in these cases physical adsorption may play a role. Also, physical adsorption serves to concentrate the molecules of a substance at a surface. This can be of importance in cases involving reaction between a chemisorbed reactant and a reactant which can be physically adsorbed. In such a system the catalytic reaction would occur between chemisorbed and physically adsorbed reactants.

The amount of physical adsorption decreases rapidly as the temperature is raised and is generally very small above the critical temperatures of the adsorbed component. This is further evidence that physical adsorption is not responsible for catalysis. For example, the rate of oxidation of sulfur dioxide on a platinum catalyst becomes appreciable only above 300℃; yet this is considerably above the critical temperature of sulfur dioxide (157℃) or of oxygen (-119℃). Physical adsorption is not highly dependent on the irregularities in the nature of the surface, but is usually directly proportional to the amount of surface. However, the extent of adsorption is not limited to a monomolecular layer on the solid surface, especially near the condensation temperature. As the layers of molecules build up on the solid surface, the process becomes progressively more like one of condensation.

Physical adsorption studies are valuable in determining the physical properties of solid catalysts. Thus the questions of surface area and pore-size distribution in porous catalysts can be answered from physical-adsorption measurements. These aspects of physical adsorption are considered in other section. The second type of adsorption is specific and involves forces much stronger than in physical adsorption. According to Langmuir's pioneer work, the adsorbed molecules are held to the surface by valence forces of the same type as those occurring between atoms in molecules. He observed that a stable oxide film was formed on the surface of tungsten wires in the presence of oxygen. This material was not the normal oxide WO_3, because it exhibited different chemical properties. However, analysis of the walls of the vessel holding the wire indicated that WO_3 was given off from the surface upon desorption. This suggested a process of the type

$$3WO_3 + 2W \longrightarrow 2[W \cdot O_3]$$
$$2[W \cdot O_3] \longrightarrow 2WO_3$$

where $[W \cdot O_3]$ represents the adsorbed compound. Further evidence for the theory that such adsorption involves valence bonds is found in the large heats of adsorption. Observed values are of the same magnitude as the heat of chemical reactions 0.5 to 100 kcal/g mol.

Taylor suggested the name chemisorption for describing this second type of combination of gas molecules with solid surfaces. Because of the high heat of adsorption, the energy possessed by chemisorbed molecules can be substantially different from that of the molecules alone. Hence the energy of activation for reactions involving chemisorbed molecules can be considerably less than that for reactions involving gas-phase molecules. It is on this basis that chemisorption offers an explanation for the catalytic effect of solid surfaces.

Two kinds of chemisorption are encountered: activated and less frequently, nonactivated. Activated chemisorption means that the rate varies with temperature according to a finite activation energy in the Arrhenius equation. However, in some systems chemisorption occurs very rapidly, suggesting an activation energy near zero. This is termed nonactivated chemisorption. It is often found that for a given gas and solid the initial chemisorption is nonactivated, while later stages of the process are slow and temperature dependent (activated adsorption).

An approximate, qualitative relation between temperature and quantity adsorbed (both physically and chemically) is illustrated as follow. Chemisorption is assumed to be activated in this case. When the critical temperature of the component is exceeded, physical adsorption approaches a very low equilibrium value. As the temperauter is raised, the amount of activated adsorption becomes important because the rate is high enough for significant quantity to be adsorbed in a reasonable amount of time. In an ordinary adsorption experiment involving the usual time periods the adsorption curve actually rises with increasing temperatures from the minimum value. When the temperature is increased still further, the decreasing equilibrium value for activated adsorption retards the process, and the quantity adsorbed passes through a maximum. At these high temperatures even the rate of the relatively slow activated process may be sufficient to give results closely approaching equilibrium. Hence the solid curve representing the amount adsorbed approaches the dashed equilibrium value for the activated adsorption process.

It has been explained that the effectiveness of solid catalysts for reactions of stable molecules is dependent upon chemisorption. Granting this, the temperature range over which a given catalyst is effective must coincide with the range where chemisorption of one or more of the reactants is appreciable. There is a relationship between the extent of chemisorption of a gas on a solid and the effectiveness of the solid as a catalyst. For example, many metallic and metal-oxide surfaces adsorb oxygen easily, and these materials are also found to be good catalysts for oxidation reactions. When reactions proceed catalytically at low temperatures, in these cases catalysis is due to nonactivated chemisorption. Thus ethylene is hydrogenated on nickel at $-78\,℃$, at which temperature there would surely exist physical adsorption of the ethylene.

An important feature of chemisorption is that its magnitude will not exceed that corresponding to a monomolecular layer. This limitation is due to the fact that the valence forces holding the molecules on the surface diminish rapidly with distance. These forces become too small to form the adsorption compound when the distance from the surface is much greater than usual bond distances.

The differences between chemisorption and physical adsorption are summarized in Table 12-1.

Table 12-1 Physical vs. chemical adsorption

Parameter	Physical adsorption	Chemisorption
Adsorbent	All solids	Some solids
Adsorbate	All gases below critical temperature	Some chemically reactive gases
Temperature range	Low temperature	Generally high temperature
Heat of adsorption	Low($\approx \Delta H_{cond}$)	High order of heat of reaction
Rate activation energy	Very rapid, Low E	Nonactivated, Low E, activated, high, E
Coverage	Multilayer possible	Monolayer
Reversibility	Highly reversible	Often irreversible
Importance	For determination of surface area and pore size	For determination of active-center area and elucidation of surface reaction kinetics

In order to develop rate equation for catalytic reaction quantitative expression for adsorption are necessary. Langmuir proposed simple formulations for rates of adsorption and desorption of gases (applicable also to liquids) on solid surfaces. While his concern was with chemisorption, the concepts have been used for deriving a valuable relationship between the volume of a gas physically adsorbed and the surface area of the solid. Applied to chemisorption, the Langmuir treatment provides a systematic basis for developing rates of heterogeneous catalytic reactions and we use this approach in other chapter. Adsorption and desorption steps in the overall conversion of reactants to products (see list at beginning of this chapter) may be very fast and hence, occur at near equilibrium. Therefore, we need also to know the equilibrium relation between the concentration of adsorbent in the gas (or liquid) and the amount adsorbed. Such equilibrium isotherms, Langmuir, other forms and the rates of adsorption are discussed in other Section.

Selected from "Specialized Chemical English (unformal published) (volumn two), by Cui Bo etc., 1994, 112-119"

Exercise

Put the following into Chinese:
   susceptible   reduction   isotherm   crystal
   chemisorption   vibrational   condensation   activity
   substantially   equilibrium constant   rotational

Reading Material

Adsorption Isotherms

The Langmuir Treatment of Adsorption. The derivations may be carried out by using as a measure of amount adsorbed either the fraction of the surface covered or the concentration of the gas adsorbed on the surface. Both procedures will be illustrated. Although the second is the more useful for kinetic developments. The important assumptions are as follows.

1. All the surface of the catalyst has the same activity for adsorption: i.e., it is energetically uniform. The concept of nonuniform surface with active centers can be employed if it is assumed that all the active centers have the same activity for adsorption and that the rest of the surface has none, or that an average activity can be used.

2. There is no interaction between adsorbed molecules. This means that the amount adsorption has no effect on the rate of adsorption per site.

3. All the adsorption occurs by the same mechanism, and each adsorbed complex has the same structure.

4. The extent of adsorption is less than one complete monomolecular layer on the surface.

In the system of solid surface and gas, the molecules of gas will be continually striking the surface and a fraction of these will adhere. However, because of their kinetic, rotational, and vibrational energy, the more energetic molecules will be continually leaving the surface. An equilibrium will be established such that the rate at which molecules strike the surface, and remain for and appreciable length of time, will be exactly balanced by the rate at which molecules leave the surface.

The rate of adsorption r_a will be equal to the rate of collision r_c of molecules with the surface multiplied by a factor F representing the fraction of the colliding molecules that adhere. At a fixed temperature the number of collisions will be proportional to the pressure p of the gas (or its concentration), and the fraction F will be constant. Hence the rate of adsorption per unit of bare surface will be $r_c F$. This is equal to kp, where k is a constant involving the fraction F and the proportionality between r_c and p.

Since the adsorption is limited to complete coverage by a monomolecular layer, the surface may be divided into two parts: the fraction θ covered by the adsorbed molecules and the fraction $1-\theta$, which is bare. Since only those molecules striking the uncovered part of the surface can be adsorbed, the rate of adsorption per unit of total surface will be proportional to $1-\theta$, that is,

$$r_a = kp(1-\theta) \qquad (12\text{-}1)$$

The rate of desorption will be proportional to the fraction of covered surface

$$r_d = k'\theta \qquad (12\text{-}2)$$

The amount adsorbed at equilibrium is obtained by equating r_a and r_d and solving for θ. The result, called the Langmuir isotherm, is

$$\theta = \frac{kp}{1+kp} \qquad (12\text{-}3)$$

where $K=k/k'$ is the adsorption equilibrium constant, expressed in units of (pressure)$^{-1}$. The fraction θ is proportional to volume of gas adsorbed, v, since the adsorption is less than a monomolecular layer. Hence Eq.(12-3) may be regarded as a relationship between the pressure of the gas and the volume adsorbed. This is indicated by writing $\theta=V/V_m$, where V_m is the volume adsorbed when all the active sites are covered, i.e., when there is a complete monomolecular layer.

The concentration form of Eq.(12-3) can be obtained by introducing the concept of an adsorbed concentration C, expressed in moles per gram of catalyst. If C_m represents the concentration corresponding to a complete monomolecular layer on the catalyst, then the rate of adsorption, mol/(s) (g catalyst), is by analogy with Eq.(12-1).

Selected from "Specialized Chemical English (unformal published) (volumn two), by Cui Bo etc., 1994, 119-122"

Lesson 13
Filtration

The separation of solids from a suspension in liquid by means of a porous medium[1] or screen which retains the solids and allows the liquid to pass is termed filtration.

In general, the pores of the medium will be larger than the particles which are to be removed, and the filter will work efficiently only after an initial deposit has been trapped in the medium. In the chemical laboratory, filtration is often carried out in a form of Buchner funnel[2], and the liquid is sucked through the thin layer of particles using a source of vacuum: in even simpler cases the suspension is poured into a conical funnel fitted with a filter paper[3]. In the industrial equivalent of such an operation, difficulties are involved in the mechanical handling of much large quantities of suspension and solids. A thicker layer of solids has to form and, in order to achieve a high rate of passage of liquid through the solids higher pressures will be needed, and it will be necessary to provide a far greater area. A typical filtration operation shows the filter medium, its support and the layer of solids, or filter cake, which has already formed.

The volumes of the suspension to be handled will vary from the extremely large quantities involved in water purification and ore handling in the mining industry to relatively small quantities in the fine chemical industry where the variety of solids will be considerable[4]. In most instances in the chemical industry it is the solids that are required and their physical size and properties are of paramount importance. Thus the main factors to be considered when selecting equipment and operating conditions are:

  a) The properties of the liquid, particularly its viscosity, density and corrosive properties.
  b) The nature of the solid—its particle size and shape, size distribution, and packing characteristics.
  c) The concentration of solids in suspension.
  d) The quantity of material to be handled, and its value.
  e) Whether the valuable product is the solid, the fluid, or both.
  f) Whether it is necessary to wash the filtered solids.
  g) Whether very slight contamination caused by contact of the suspension or filtrate with the various components of the equipment is detrimental to the product.
  h) Whether the feed liquor may be heated.
  i) Whether any form of pretreatment would be helpful.

Filtration is essentially a mechanical operation and is less demanding in energy than evaporation or drying where the high latent heat[5] of the liquid, which is usually water, has to be provided. In the typical operation the cake gradually builds up on the medium and the resistance to flow progressively increases. During the initial period of flow, particles are deposited in the surface layers the cloth to form the true filtering medium. This initial deposit may be formed from a special initial flow of precoat material. The most important factors on which the rate of filtration then depends will be:

a) The drop in pressure from the feed to the far side of the filter medium.
b) The area of the filtering surface.
c) The viscosity of the filtrate.
d) The resistance of the filter cake.
e) The resistance of the filter medium and initial layers of cake.

The type of filtration described above is usually referred to as cake filtration[6] the proportion of solids in the suspension is large and most of the particles are collected in the filter cake which can subsequently detached from the medium. Where the proportion of solids is very small, as for example in air or water filtration, the particles will often be considerably smaller than the pores of the filter medium and will penetrate a considerable depth before being captured; such a process is called deep bed filtration[7]. In the reading material the types of filtration equipment will be considered.

Selected form "English for Chemical Engineers, by Ma Zhengfei etc., Southeast University Press, 2006, 25-27"

New Words

1. suspension [səˈspenʃən]   n. 悬浮液
2. screen [skri:n]   n. 筛，网
3. suck [sʌk]   v. 吸入
4. paramount [ˈpærəmaunt]   a. 首要的, 最高的
5. corrosive [kəˈrəusiv]   a. 腐蚀的
6. contamination [kəntæmiˈneiʃən]   n. 污染，污物
7. detrimental [ˌdetriˈmentl]   a. 有害的，不利的
8. subsequently [ˈsʌbsikwəntli]   ad. 其后，其次
9. detached [diˈtætʃt]   a. 分离的，孤立的

Notes

1. porous medium   多孔介质。
2. Buchner funnel   布氏漏斗。
3. filter paper   滤纸。
4. The volumes of the suspension to be handled will vary from the extremely large quantities involved in water purification and ore handling in the mining industry to relatively small quantities in the fine chemical industry where the variety of solids will be considerable. 参考译文：要处理的悬浮液量变化会很大，从处理量很大的水净化和采矿工业中的矿物到精细化工中处理量相对较小的涉及各种固体的情况。
5. latent heat   潜热。
6. cake filtration   滤饼。
7. Such a process is called deep bed filtration. 参考译文：这一过程称为深床过滤。

Exercise

1. Put the following into Chinese:

Buchner funnel   trap   subsequently
screen   viscosity   liquor

| deposit | contamination | detached |

2. Put the following into English:

多孔介质	滤纸	潜热
悬浮液	滤饼	深床过滤
板框式压滤机	有害的	滤液

3. Comprehension and toward interpretation
   a. What is filtration?
   b. What is the difference of filtration between chemical laboratory and industry?
   c. Which factors should be considered in the operation of filtration?
   d. Which factors will have a close relation with the rate of filtration?

Reading Material

Types of Filtration Equipment

Classification of Filters. There are a number of ways to classify types of filtration equipment, and it is not possible to make a simple classification that includes all types of filters. In one classification filters are classified according to whether the filter cake is the desired product or whether the clarified filtrate or outlet liquid is desired. In either case the slurry can have a relatively large percentage of solids so that a cake is formed, or have just a trace of suspended particles.

Filters can be classified by operating cycle. Filters can be operated as batch, where the cake is removed after a run, or continuous, where the cake is continuously removed. In another classification, filters can be of the gravity type, where the liquid simply flows by a hydrostatic head, or pressure or vacuum can be used to increase the flow rates. An important method of classification depends upon the mechanical arrangement of the filter media. The filter cloth can be in a series arrangement as flat plates in an enclosure, as individual leaves dipped in the slurry, or on rotating-type rolls in the slurry. In the following sections only the most important types of filters will be described.

Bed Filters. The simplest type of filter is the bed filter. This type is useful mainly in cases where relatively small amounts of solids are to be removed from large amounts of water in clarifying the liquid. Often the bottom layer is composed of coarse pieces of gravel resting on a perforated or slotted plate. Above the gravel is fine sand, which acts as the actual filter medium. Water is introduced at the top onto a baffle which spreads the water out. The clarified liquid is drawn out at the bottom.

The filtration continues until the precipitate of filtered particles has clogged the sand so that the flow rate drops. Then the flow is stopped and water introduced in the reverse direction so that it flows upward, backwashing the bed and carrying the precipitated solid away. This apparatus can only be used on precipitates that do not adhere strongly to the sand and can be easily removed by backwashing. Open tank filters are used in filtering municipal water supplies.

Plate-and-Frame Filter Presses. One of the important types of filters is the plate-and-frame filter press. These filters consist of plates and frames assembled alternatively with a filter cloth over each side of the plates. The plates have channels cut in them so that clear filtrate liquid can drain down along each plate. The feed slurry is pumped into the press and flows through the duct into each of the open frames so that slurry fills the frames. The filtrate flows through the filter cloth and the solids build up as a cake on the frame side of the cloth. The filtrate flows between the filter cloth and the face of the plate through the channels to the outlet.

The filtration proceeds until the frames are completely filled with solids. In many cases the filter press will have a separate discharge to the open for each frame. Then visual inspection can be made to see if the filtrate is running clear. If one is running cloudy because of a break in the filter cloth or other factors, it can be shut off separately. When the frames are completely full, the frames and plates are separated and the cake removed. Then the filter is reassembled and the cycle is repeated.

If the cake is to be washed, the cake is left in the plates and through washing is performed. In this press a separate channel is provided for the wash water inlet. The wash water enters the inlet, which has ports opening behind the filter cloths at every other plate of the filter press. The wash water then flows through the filter cloth, through the entire cake, through the filter cloth at the other side of the frames, and out the discharge channel. It should be noted that there are two kinds of plates: those having ducts to admit wash water behind the filter cloth, alternating with those without such ducts.

The plate-and-frame presses suffer from the disadvantages common to batch processes. The cost of labor for removing the cakes and reassembling plus the cost of fixed charges for downtime can be an appreciable part of the total operating cost. Some newer types of plate-and frame presses have duplicates sets of frames mounted on a rotating shaft. Half of the frames are in use while the others are being cleaned, saving downtime and labor costs. Other advances in automation have been applied to these types of filters.

Filters presses are used in batch processes but cannot be employed for high-throughput processes. They are simple to operate, very versatile and flexible in operation, and can be used at high pressures, when necessary, if viscous solutions are being used or the filter cake has a high resistance.

Leaffilters. The filter press is useful for many purposes but is not economical for handling large quantities of sludge or for efficient washing with a small amount of wash water. The wash water often channels in the cake and large volumes of wash water may be needed. The leaf filter was developed for large volumes of slurry and more efficient washing. Each leaf is a hollow wire framework covered by a sack of filter cloth.

A number of these leaves are hung in parallel in a closed tank. The slurry enters the tank and is forced under pressure through the filter cloth, where the cake deposits on the outside of the leaf. The filtrate flows inside the hollow framework and out a header. The wash liquid follows the same path as the slurry. Hence, the washing is more efficient than the through washing in plate-and-frame filter presses. To remove the cake, the shell is opened. Sometimes air is blown in the reverse direction into the leaves to help in dislodging the cake. If the solids are not wanted, water jets can be used to simply wash away the cakes without opening the filter.

Leaf filters also suffer from the disadvantage of batch operation. They can be automated for the filtering, washing, and cleaning cycle. However, they are still cyclical and are used for batch processes and relatively modest throughput processes.

Continuous Rotary Filters. The plate-and-frame filters suffer from the disadvantages common to all batch processes and cannot be used for large capacity processes. A number of continuous-type filters are available as discussed below.

a) Continuous rotary vacuum-drum filter. This filter filters, washes, and discharges the cake in a continuous repeating sequence. The drum is covered with a suitable filtering medium. The drum rotates and an automatic valve in the center serves to activate the filtering, drying, washing, and cake discharge functions in the cycle. The filtrate leaves through the axle of the filter. The automatic valve provides separate outlets for the filtrate and the wash liquid. Also, if needed, a connection for compresses air blowback just before discharge can be used to help in cake removal by the knife scraper. The maximum pressure differential for the vacuum filter is only 1 atm. Hence, this type is not suitable for viscous liquids or for liquids that must be enclosed. If the drum is enclosed in a shell, pressures above atmospheric can be used.

b) Continuous rotary disk filter. This filter consists of concentric vertical disks mounted on a horizontal rotating shaft. The filter operates on the same principle as the vacuum rotary drum filter. Each disk is hollow and covered with a filter cloth and is partly submerged in the slurry. The cake is washed, dried, and scraped off when the disk is in the upper half of its rotation. Washing is less efficient than with a rotating drum type.

c) Continuous rotary horizontal filter. This type is vacuum filter with the rotating annular filtering surface divided into sectors. As the horizontal filter rotates it successively receives slurry, is washed, dried, and the cake is scraped off. The washing efficiency is better than with the rotary disk filter. This filter is widely used in ore extraction processes, pulp washing, and other large-capacity processes.

Selected from "English for Chemical Engineers, by Ma Zhengfei etc., Southeast University Press, 2006, 28-30"

Lesson 14
Evaporation, Crystallization and Drying

Evaporation. Heat transfer to a boiling liquid occurs so often that it is considered an individual operation. It is called evaporation. The objective of evaporation is to concentrate a solution consisting of a nonvolatile solute and a volatile solvent. In the overwhelming majority of evaporations the solvent is water. Evaporation is conducted by vaporizing a portion of the solvent to produce a concentrated solution or thick liquor. Evaporation differs from drying in the residue is a liquid — sometimes a highly viscous one — rather than a solid; it differs from distillation in that the vapor usually is a single component, and even when the vapor is a mixture, no attempt is made in the evaporation step to separated the vapor into fractions; it differs from crystallization in that emphasis is placed on concentrating a solution rather than forming and building crystals. In certain situations, e.g., in the evaporation of brine to produce common salt, the line between evaporation and crystallization is far from sharp. Evaporation sometimes produces slurry of crystals in a saturated mother liquor.

Normally, in evaporation the thick liquor is the valuable product and the vapor is condensed and discarded. In one specific situation, however, the reverse is true. Mineral — bearing water is often is evaporated to give a solid—free product for boiler feed, for special process requirements, or for human consumption. This technique is often called water distillation, but technically it is evaporation. Large-scale evaporation processes are being developed and used for recovering potable water from seawater. Here the condensed water is the desired product. Only a fraction of the total water in the feed is recovered, and the remainder is returned to the sea.

Liquid Characteristics. The practical solution of an evaporation problem is profoundly affected by the character of the liquor to be concentrated. It is the wide variation in liquor characteristics (which demands judgment and experience in engineering and operating evaporators) that broadens this operation from simple heat transfer to a separate art. Some of the most important properties of evaporating liquids are as follows.

Single and Multiple — Effect Operation. Most evaporators are heated by steam condensing on metal tubes. Nearly always the material to be evaporated flows inside the tube. Usually the steam is at low pressure, below 3 atm, often the boiling liquid is under a moderate vacuum, up to about 0.5 atm abs. Reducing the boiling temperature of the liquid increases the temperature difference between the steam and the boiling liquid and thereby increases the heat transfer rate in the evaporator.

When a single evaporator is used, the vapor from the boiling liquid is condensed and discarded. This method is called single — effect evaporation, and although it is simple, it utilizes steam ineffectively. To evaporate 1 b of water from a solution calls for from 1 to 1.3 b of steam. If the vapor from one evaporator is fed into the steam chest of a second evaporator and the vapor from the second is then sent to a condenser, the operation becomes double — effect. The heat in the original steam is reused in the second effect, and the evaporation achieved by a unit mass of steam fed to the first effect is approximately doubled.

Additional effects can be added in the same manner. The general method of increasing the evaporation per pound of steam by using a series of evaporators between the steam supply and the condenser is called multiple — effect evaporation.

Crystallization. Crystallization from liquid solution is important industrially because of the variety of materials that are marketed in the crystalline form. Its wide use is based on the fact that a crystal formed from an important solution is itself pure (unless mixed crystals occur) and that is crystallization affords a practical method of obtaining pure chemical substances in a satisfactory condition for package and storing.

It is clear that good yield and high purity are important objectives in operating a crystallization process, but these two factors are not the only ones to be considered. The appearance and size range of a crystalline product are also significant. It is especially necessary that the crystals should be of reasonable and uniform size. If they are to be further processed, reasonable size and uniformity are desirable for washing, filtering, reacting with other chemicals, transporting, and storing the crystals. If the crystals are to be marketed as a final product, customers require individual crystals to be strong, nonaggregated, uniform in size, and noncaking in the package. For these reasons, crystal size distribution (CSD) must be under control; it is a prime objective in the design and operation of crystallizers.

In general, crystallization may be analyzed from the standpoint of purity, yield, energy requirements, and rates of formation and growth of crystals.

Drying. General, drying a solid means the removal of relatively small amounts of water or other liquid from the solid material to reduce the content of residual liquid to an acceptably low value. Drying is usually the final step in a series of operations, and the product from a dryer is often ready for final packaging.

Water or other liquids may be removed from solids mechanically by pressers or centrifuger, thermally by vaporization. This chapter is restricted to drying by thermal vaporization. It is generally cheaper to remove water mechanically than thermally, and thus it is advisable to reduce the moisture content as much as practicable before feeding the material to a heated dryer.

The moisture content of a dried substance varies from product to product. Occasionally the product contains no water, and is called bone-dry. More commonly, the product does contain some water. Dried table salt, for example, contains 0.5 percent water, dried coal about 4 percent, and dried casein about 8 percent. Drying is a relative term and means merely that there is a reduction in moisture content from a initial value to some acceptable final value.

Selected from "English for Chemistry and Chemical Engineering, by Zhang Yuping etc., Chemical Industry Press, 2007, 92-93"

New Words

1. market ['mɑːkit]　*n.* 市场，菜市场；*vt.* （在市场上）销售
2. crystalline ['kristəlain]　*a.* 结晶（质）的，结晶状的
3. be based on　以……为依据，基于……
4. be formed from　由……生成；由运输、输送、运送形成
5. impure [im'pjuə]　*a.* 不纯的，有杂质的
6. crystallization [kristəlai'zeiʃən]　*n.* 结晶
7. afford [ə'fɔːd]　*vt.* 担负得起（费用、损失等），抽得出（时间），提供
8. package ['pækidʒ]　*n.* 包装，打包；（中、小型的）包裹，包；（商品、产品等的）一件
9. store [stɔː]　*vt.* 储藏，储备；装备

10. purity ['pjuəriti]   n. 纯度，纯洁
11. factor ['fæktə]   n. 因素，要素，因子
12. appearance [ə'piərəns]   n. 外貌，外表，出现，露面
13. significant [sig'nifikənt]   a. 有意义的，意味深长的，重要的
14. uniformity [,ju:ni'fɔ:miti]   n. 一致性，均匀性，均一性
15. desirable [di'zairəbl]   a. 合意的，希望得到的，合乎需要的
16. wash [wɔʃ]   v. 洗涤，洗
17. transport [træns'pɔ:t]   vt. 运输，输送，运送；n. 运输，输送，运送
18. be + 不定式或其短语   拟将……，准备……
19. final ['fainl]   a. 最后的，最终的
    a final product   最终产品
20. customer ['kʌstəmə]   n. 顾客，主顾；用户
21. individual [,indi'vidjuəl]   a. 个人的，个别的，单独的
22. nonaggregate [,nɔn'ægrigit]   v. 不聚集，不聚结成团
23. noncake ['nɔn'keik]   v. 不结成块，使……不结块
24. crystal size distribution （缩写为CSD）晶体大小的分配
25. be under control   受到控制，受到支配
26. prime [praim]   a. 最初的，基本的
27. analyze ['ænəlaiz]   vt. 分析，分解
28. standpoint ['stændpɔint]   n. 立场，观点
    from the standpoint of   ……站在……的立场上；根据……的观点
29. growth [grəuθ]   v. 成长，生长；长大
30. packing   n. 包装，打包
31. thermally   adv. 用热的方法
32. restrict [ris'trikt]   v. 局限，限制性
33. occasionally [ə'keiʒənəli]   adv. 偶尔，有时，间或
34. bone-dry   a. 干透的
35. casein ['keisiin]   n. 酪素，酪蛋白
36. granular ['grænjulə]   a. 粒状的，晶粒的

Exercise

1. Put the following into Chinese:

   crystallization   appearance   granular   free flowing
   impure   customer   calandria   supersaturation
   magma   prime   solar pan   spray dryer

2. Put the following into English：

   结晶的   因素   分析   刮膜式蒸发器
   包装   均匀性   观点   短管蒸发器
   纯度   运输   直接加热蒸发器   气流干燥器

Reading Material

Evaporator, Crystallizer, and Dryer

1. Evaporator

A great variety of evaporator designs have been developed for specialized applications in particular industries. The designs can be grouped into the following basic types.

Direct Heated Evaporators. This type includes solar pans and submerged combustion units. Submerged combustion evaporators can be used for applications where contamination of the solution by the products of combustion is acceptable.

Long-Tube Evaporators. In this type the liquid flows as a thin film on the walls of a long, vertical, heated tube. Both falling film and rising film types are used. They are high capacity units; suitable for low viscosity solutions.

Forced Circulation Evaporators. In forced circulation evaporators the liquid is pumped through the tubes. They are suitable for use with materials which tend to foul the heat transfer surfaces, and where crystallization can occur in the evaporator.

Agitated Thin-Film evaporators. In this design a thin layer of solution is spread on the heating surface by mechanical means. Wiped-film evaporators are used for very viscous materials and for producing solid products.

Short-Tube Evaporators. Short-tube evaporators also called calandria evaporators, are used in the sugar industry.

Evaporator Selection. The selection of the most suitable evaporator type for a particular application will depend on the following factors:
(1) The throughput required.
(2) The viscosity of the feed and the increase in viscosity during evaporation.
(3) The nature of the product required: solid, slurry, or concentrated solution.
(4) The heat sensitivity of the product.
(5) Whether the materials are fouling or non-fouling.
(6) Whether the solution is likely to foam.
(7) Whether direct heating can be used.

Auxiliary Equipment. Condensers and vacuum pumps will be needed for evaporators operated under vacuum. For aqueous solutions, steam ejectors and jet condensers are normally used. Jet condensers are direct-contact condensers, where the vapor is condensed by contact with jets of cooling water.

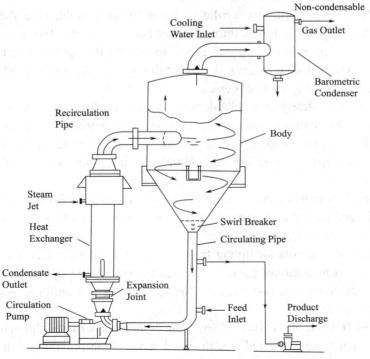

Fig.14-1 Circulating magma crystallizer (evaporative type)

2. Crystallizer

Crystallization equipment can be classified by the method used to obtain supersaturation of the liquor, and also by the method used to suspend the growing crystals. Supersaturation is obtained by cooling or evaporation. There are four basic types of crystallizer: tank crystallizers, scraped-surface crystallizers, circulating magma crystallizers (Fig. 14-1) and circulating liquor crystallizers.

Typical applications of the main types of crystallizer are summarized in Table 14-1.

Table 14-1 Selection of crystallizers

Crystallizer type	Applications	Typical uses
Tank	Batch operation, small-scale production	Fatty acids, vegetable oils, sugars
Scraped surface	Organic compounds, where fouling is a problem, viscous materials	Chlorobenzenes, organic acids, paraffin waxes, naphthalene, urea
Circulating magma	Production of large-sized crystals. High throughputs	Ammonium and other inorganic salts, sodium and potassium chlorides
Circulating liquor	Production of uniform crystals (smaller size than circulating magma). High throughputs.	Gypsum, inorganic salts, sodium and potassium nitrates, silver nitrates.

3. Dryer

The basic types used in the chemical process industries are:

Tray Dryers. Batch tray dryers are used for drying small quantities of solids, and are used for a wide range of materials. The material to be dried is placed in solid bottomed trays over which hot air is blown; or perforated bottom trays through which the air passes. Batch dryers have high labor requirements, but close control can be maintained over the drying conditions and the product inventory, and they are suitable for drying valuable products.

Conveyor Dryers (Continuous Circulation Band Dryers). In this type, the solids are fed on to an endless, perforated, conveyor belt, through which hot air is forced. The belt is housed in a long rectangular cabinet, which is divided up into zones, so that the flow pattern and temperature of the drying air can be controlled. The relative movement through the dryer of the solids and drying air can be parallel or, more usually, counter-current.

This type of dryer is clearly only suitable for materials that form a bed with an open structure. High drying rates can be achieved, with good product-quality control. Thermal efficiencies are high and, with steam heating, steam usage can be as low as 1.5 kg per kg of water evaporated. The disadvantages of this type of dryer are high initial cost and, due to the mechanical belt, high maintenance costs.

Rotary Dryer. In rotary dryers the solids are conveyed along the inside of a rotating, inclined, cylinder and are heated and dried by direct contact with hot air gases flowing through the cylinder. In some, the cylinders are indirectly heated.

Rotating dryers are suitable for drying free-flowing granular materials. They are suitable for continuous operation at high throughputs; have a high thermal efficiency and relatively low capital cost and labor costs. Some disadvantages of this type are: a non-uniform residence time, dust generation and high noise levels.

Fluidized Bed Dryers. In this type of dryer, the drying gas is passed through the bed of solids at a velocity sufficient to keep the bed in a fluidized state; which promotes high heat transfer and drying rates. Fluidized bed dryers are suitable for granular and crystalline materials within the particle size range 1 to 3 mm. They are designed for continuous and batch operation. The main advantages of fluidized dryers are: rapid and uniform heat transfer, short drying times, with good control of the drying conditions; and low floor area requirements. The power requirements are high compared with other types.

Pneumatic Dryers. Pneumatic dryers also called flash dryers, are similar in their operating principle to spray dryers. The product to be dried is dispersed into an upward-flowing stream of hot gas by a suitable feeder. The equipment acts as a pneumatic conveyor and dryer. Contact times are short, and this limits the size of particle-that can be dried. Pneumatic dryers are suitable for materials that are too fine to be dried in a fluidized bed dryer but which are heat sensitive and must be dried rapidly. The thermal efficiency of this type is generally low.

Spray Dryers. Spray dryers are normally used for liquid and dilute slurry feeds, but can be designed to handle any material that can be pumped. The material to be dried is atomized in a nozzle, or by a disc-type atomizer, positioned at the top of a vertical cylindrical vessel. Hot air flows up the vessel (in some designs downward) and conveys and dries the droplets. The liquid vaporizes rapidly from the droplet surface and open, porous particles are formed. The dried particles are removed in a cyclone separator or bag filter.

The main advantages of spray drying are the short contact time, making it suitable for drying heat-sensitive materials, and good control of the product particle size, bulk density, and form. Because the solids concentration in the feed is low, the heating requirement will be high.

Rotary Drum Dryers. Drum dryers are used for liquid and dilute slurry feeds. They are an alternative choice to spray dryers when the material to be dried will form a film on a heated surface, and is not heat sensitive.

Selected from "Chemical Engineering, Vol. 6, 2nd Edition, R. K. Sinnott, Pergamon Press, 1996"

New Words

1. calandria [kə'lændriə] *n.* 排管式，加热管群
2. direct heated evaporator 直接加热蒸发器
3. solar pan 盐池，盐田
4. submerged combustion 浸没燃烧
5. agitated thin-film evaporator 搅拌式薄膜蒸发器
6. wiped-film evaporator 刮膜式蒸发器
7. short tube evaporator 短管蒸发器
8. steam ejector 蒸汽喷射器
9. jet condenser 喷射式冷凝器
10. free flowing 自由流动的，流动性能良好的
11. supersaturation ['sju:pə'sætə'reiʃən] *n.* 过饱和（现象）
12. tank crystallizer 槽式结晶器
13. scraped-surface crystallizer 刮膜式结晶器
14. magma ['mægmə] *n.* 稀糊状混合物，岩浆，稠液
15. circulating magma crystallizer 晶浆循环结晶器
16. circulating liquor crystallizer 母液循环结晶器
17. tray dryer 盘架干燥器
18. conveyor dryer 带式干燥器
19. continuous circulation band dryer 连续循环带式干燥器
20. cabinet ['kæbinit] *n.* 箱，室，壳体
21. fluidized bed dryer 流化床干燥器
22. pneumatic dryer 气流干燥器
23. flash dryer 气流干燥器
24. spray dryer 喷雾干燥器
25. pneumatic [nju:'mætik] *a.* 气力的，气动的
26. bulk density 堆积密度
27. rotary drum dryer 转鼓式干燥器

Lesson 15
Computer-Assisted Design of New Process

Designs for new processes proceed through at least three stages:

Conceptual design[1] — the generation of ideas for new processes (process synthesis) and their translation into an initial design. This stage includes preliminary cost estimates to assess the potential profitability of the process, as well as analyses of process safety and environmental considerations.

Final design[2] —a rigorous set of design calculations to specify all the significant details of a process.

Detailed design[3]—preparation of engineering drawing and equipment lists needed for construction.

The key step in the conceptual design of a new chemical manufacturing process is generating the process flowsheet[4]. All other elements of computer-aided design (e.g., process simulation, design of control systems, and plantwide integration of processes[5]) come into play[6] after the flowsheet has been established. In current practice, the pressure to enter the market quickly often allows for the exploration of only a few of the process alternatives that should be considered[7]. To be fair to today's designers, it is possible to generate a very large number of alternative process paths at the conceptual stage of design, and yet experience indicates that less than 1 percent of the ideas for new designs become commercial. Thus, the challenge in computer-aided process synthesis is to develop systematic procedures for the generation and quick screening of many process alternatives. The goal is to simplify the synthesis/analysis activity in conceptual design and give the designer confidence that the initial universe of potential process paths contained all the pathways with reasonable chances for commercial success[8]. The advances in computer-aided process synthesis that are possible over the next decade are dramatic. They include both an increasing level of sophistication (e.g., the synthesis of heat exchanger networks, sequences of separation processes, networks of reactors, and process control systems) and computational procedures that should make possible option in a relatively short amount of time.

As the designer moves from conceptual design toward final design, he or she must analyze a number of alternatives for the final design. The development of large, computer-aided design programs (so-called process simulators) such as FLOWTRAN, PROCESS, DESIGN 2000, and ASPEN (or other equivalent programs used in various companies) has significantly automated the detailed computations needed to analyze these various process designs[9]. The availability of process simulators has probably been the most important development in the design of petrochemical plants in the past 20 years, cutting design times drastically and resulting in better designed plants.

Although the available simulators have done much to achieve superior design of petrochemical processes, there is considerable room for improvement. For example, better models are needed for complex reactors and for solids processing operations such as crystallization, filtration, and drying. Thermodynamic models are needed for polar compounds. Moreover, the current process simulators are limited to steady-state operations and are capable of analyzing only isolated parts of a chemical plant at any given time. This compartmentalization is due to the limitations on computer memory that prevailed when these programs were first developed. This memory limitation resulted in a computational strategy that divided the plant into "boxes" and simulated static conditions within each box, iteratively merging the results to simulate the entire plant. With today's supercomputers, it is possible to simulate the dynamics of the entire chemical plant. This opens the way for dramatic advances in modeling and analysis of alternative process designs, because the chemical reactions that occur in manufacturing process are usually nonlinear and interdependent, and random disturbances in the process can propagate quickly and threaten the operation of the entire plant. To nullify the effects of such disturbances, the designer must know the dynamics of the entire plant, so that control failure in any one unit does not radiate quickly to other units. It is now within our reach to integrate this sophisticated level of design and analysis on a plantwide scale (including design and performance modeling of plantwide control systems) into the computational tools used to analyze and optimize the performance of individual processes in the plant.

In the detailed design stage for a chemical manufacturing process, a chosen process design must be converted into a list of equipment items to be purchased and a set of blueprints to guide their assembly. The design is presented as a detailed process flow diagram (PFD)[10], from which is constructed a list of all needed items of equipment, and piping and instrumentation diagrams (PIDs)[11] that show the equipment and its interconnections. The next task is to establish the physical layout for the entire plant. Advanced computer-based drafting tools aid in all these activities.

Selected from "English for Chemical Engineers, by Ma Zhengfei etc., Southeast University Press, 2006, 105-108"

New Words

1. generation [dʒenə'reiʃən] n. 产生，形成，引起
2. translation [træns'leiʃən] n. 转化，变换
3. preliminary [pri'liminəri] a. 初始的，初步的
4. exploration [eksplɔ:'reiʃən] n. 探究，探索，探测
5. screening ['skri:niŋ] n. 筛选，甄别
6. pathway ['pɑ:θwei] n. 路径，通道，轨迹
7. sophistication [səfisti'keiʃən] n. 完善，改进，复杂化
8. network ['netwə:k] n. 网络
9. automate ['ɔ:təmeit] v. 使……自动化
10. simulator ['simjuleitə] n. 模拟器
11. petrochemical [,petrəu'kemikl] a. 石油化学的
12. crystallization [kristəlai'zeiʃn] n. 结晶
13. filtration [fil'treiʃən] n. 过滤
14. drying ['draiiŋ] n. 干燥
15. polar ['pəulə] a. 极性的
16. compartmentalization [kɔmpɑ:tmentəlai'zeiʃən] n. 区域化，隔开
17. merge [mə:dʒ] v. 合并，汇合
18. interdependent [intədi'pendənt] a. 互相依赖的，互相影响的
19. propagate ['prɔpəgeit] v. 传播，蔓延
20. nullify ['nʌlifai] v. 废除，使……无效
21. blueprint [blu:print] n. 蓝图，设计图
22. interconnection [intə(:)kə'nekʃən] n. 互相联系，互连
23. layout ['leiaut] n. 草图，布置图，布置，安排
24. assembly [ə'sembli] n. 装配，组合

Notes

1. conceptual design 概念设计，方案设计。
2. final design 最终设计。
3. detailed design 详细设计。
4. process flowsheet 工艺流程图。
5. plantwide integration of processes 全厂范围的过程综合。
6. come into play 开始运行。
7. In current practice, the pressure to enter the market quickly often allows for the exploration of only a few of the process alternatives that should be considered. 参考译文：按目前实际做法，迅速进入市场的压力常常仅允许考察和比较部分可供选择的方案。
8. The goal is to simplify the synthesis/analysis activity in conceptual design and give the designer confidence that the initial universe of potential process paths contained all the pathways with reasonable chances for commercial success. 参考译文：目标是简化概念设计中的综合或分析工作，并使设计者对有商业成功机会的所有可能的过程流程充满信心。
9. The development of large, computer-aided design programs (so-called process simulators) such as FLOWTRAN, PROCESS, DESIGN 2000, and ASPEN (or other equivalent programs used in various companies) has significantly automated the detailed computations needed to analyze these various process designs. 参考译文：如 FLOWTRAN、PROCESS、DESIGN 2000 和 ASPEN（或各公司使用的其他相当程序）等大型计算机辅助设计程序（所谓的过程模拟器）的开发，已能有效地自动对需分析的各过程设计进行详细计算。FLOWTRAN、PROCESS、DESIGN 2000 和 ASPEN 是几个目前国际上较著名的计算机化工流程稳态模拟系统。
10. process flow diagram 过程流程图。
11. piping and instrumentation diagrams 管道布置和仪表配置图。

Exercise

1. Put the following into Chinese:

generation	screening	blueprint	schedule
preliminary	automate	interconnection	simulation
rigorous	simulator	graphical	comprehensive

2. Put the following into English:

方案设计	详细设计	过程综合	区域化
最终设计	工艺流程图	轨迹	布置图
转化	计算机辅助设计	石油化学的	简化

3. Comprehension and toward interpretation

   a. What is the conceptual design?
   b. Where one can see the advantages of computer-aided design?
   c. The available simulators have done much to achieve superior design of petrochemical processes. Can give the reasons?
   d. What is the meaning of the detailed design?

Reading Material

Process Design in the 21st Century

It is April 2008. A process engineer for a medium-size chemical company is designing a process for a new series of products her company is developing. By talking to her computer and pointing to the screen, she indicates that the processes need a reactor and some separators, as well as recycle capability. The computer, using an artificial intelligence program tied to mathematical models, determines how the reactor temperature and pressure will influence the product yield and how the yield in turn will influence the amount of process stream recycle. The computer chooses the optimal types of separators and indicates how they should be sequenced for different process streams and products of series. After 20 minutes on a teraflop machine (equivalent to 1 week of computing on a CRAY I of the 1980s), the computer indicates that the general process design is completed, having given partial information during the 20-minute computation.

The graphical output from the computer shows the process flowsheet, with several separation units and projected equipment and operating costs. It also flags information that is uncertain because it had to use thermodynamic data extrapolated from measured values. At the engineer's request, the computer shows several alternative flowsheets it had considered, indicates their projected costs, and tells why it eliminated each of them. Some of the flowsheets were eliminated because of high cost, others because they were considered unsafe, others because the startup procedures would be difficult, and still others because they were based on uncertain extrapolation of experimental data.

After perusal of these process options, the engineer asks the computer to select five designs for further study, and the computer produces a paper copy of the flowsheet and design parameters for each. She then tells the computer to prepare more rigorous designs from each of the five flowsheets. The computer questions the third design because it involves extensive extrapolation of known thermodynamic data. However, since the engineer feels that this may be one of the better designs, she asks the computer to notify the laboratory management computer, which is connected to a laboratory technician's terminal, to begin experimental determination of the missing thermodynamic data so that the design can be prepared. The computer reminds the engineer that precise control of process conditions will be essential for another of the designs and that a new sensor may be necessary. The engineer consults the artificial intelligence program, which tells her that either of two sensors being developed in another company department may fill her need. The engineer uses her computer to schedule a meeting with the manager of sensor development.

Two years later, the process is producing one of the new products. The time from design to operation was short enough that the company was able to capture a new market, and it plans to extend the business with related products of the series. The design engineer has developed a real-time dynamic simulation of the process and is running it on the company's parallel supercomputer to test strategies for producing the related products from the process equipment. She queries another computer to search the historical records of plant operation. She wants to know how well the dynamic model mirrors the actual plant operation and when the process parameters were last adjusted to gives a better fit between the plant and the model. She feeds this information to her computer, which devises a comprehensive plan for scheduling production among four of the new products, with provision for revising the schedule as inventories and sale of the changing products. A new business has been created.

Selected from "English for Chemical Engineers, by Ma Zhengfei etc., Southeast University Press, 2006, 109-110"

Lesson 16
Catalysis

When a chemist considers a chemical reaction, he generally asks himself three questions: How fast is it? How complete is it? How selection is it? Some reactions are very fast and go to completion to yield a single product. A familiar example is the reaction of sodium and chlorine to form sodium chloride. Other reactions, for instance the reaction of hydrogen and oxygen to form water, go very slowly at room temperature but are extremely fast at higher temperatures. They eventually go to completion to yield a single product. Most reactions are very slow indeed. The chemist has to find ways to speed them up. If he is lucky, he can do that simply by raising the temperature (as in the reaction of hydrogen with oxygen). Unfortunately increasing the temperature frequently produces undesirable side effects.

Two major reasons for seeking alternative means of speeding up chemical reactions can be illustrated by the following examples. One is the reaction of nitrogen and hydrogen to form ammonia (NH_3). The other is a hypothetical reaction between methane (CH_4) and oxygen to form dimethyl ether (CH_3OCH_3) and water. Both reactions are alike in that at room temperature their rate is essentially zero. The laws of thermodynamics tell us, however, that both reactions should go a long way toward completion at room temperature.

Let us see what happens as we raise the temperature with each of these reactions. In the reaction forming ammonia there is a composition (some specific proportion of H_2, N_2 and NH_3) that is in chemical equilibrium at a given temperature and pressure. Other compositions, if they react, must tend toward the equilibrium composition. Suppose one raises the temperature in an attempt to get a reasonably good yield of ammonia in a reasonable period of time. Before the rate of formation of ammonia becomes fast, the position of equilibrium has shifted to increase the proportion of hydrogen and nitrogen at the presence of ammonia. For practical purposes, then, one cannot cause nitrogen and hydrogen to combine directly to form ammonia.

In the second reaction other oxidation reactions of methane become fast as the temperature is raised and destroy the methane before detectable amounts of dimethyl ether are formed. In general most possible reactions proceed slowly at room temperature, and the great majority of reactions in this class cannot be practically affected by raising the temperature. For such reactions what one lacks is selectivity.

If nature provided no way to accelerate chemical reactions selectively, our modern technological society could not have arisen, but one must quickly add that its absence would not be noticed because no form of life could exist either. In fact, nature long ago discovered how to effect many reactions of the type that cannot be affected merely by raising the temperature. Man has acquired the knack only recently.

If the mixture of hydrogen and oxygen is exposed to plantinum powder at room temperature, a rapid reaction forming water occurs on the surface of the metal particles. A few atoms of platinum can lead to the formation of many molecules of water. This is catalysis, defined as the phenomenon in which a relatively small amount of foreign material, called a catalyst, augments the rate of a chemical reaction without itself being consumed. The chemist now has an alternate means at his disposal for speeding up reactions. How do these catalysts work?

Suppose one knows that two chemical substances, A and B, react to form C but that the reaction is extremely slow at room temperature. One can demonstrate that no combination of elementary processes involving A, B and C will result in rapid formation of C. Now add a catalyst, designated Cat. It provides the possibility of new elementary processes. If processes such as the following are fast, C will be formed rapidly:

$$A + Cat \longrightarrow ACat$$
$$ACat + B \longrightarrow C + Cat$$

The combining tendency of A and Cat must be adequate to yield the complex ACat, yet it must be not so strong as to make ACat unreactive. After all, unless ACat reacts rapidly with B to form C and then regenerates Cat, we do not have catalysis.

Chemists must by now have discovered the great majority of reactions that proceed without catalysts. Clearly the future of preparative chemistry will heavily involve catalysis. We already have a catalyst (iron) that enables us to manufacture ammonia, but we know of no catalyst that leads to the formation of dimethyl ether from methane and oxygen. There are almost innumerable other reactions that have favorable positions of equilibrium but for which no catalysts are known.

Catalytic reactions can be classified into three principal types. The most common is heterogeneous catalysis, in which the catalyst is a solid and the reactants and products are either gases or liquids. Platinum is a heterogeneous catalyst for the reaction between hydrogen and oxygen. The second type is homogeneous catalysis, in which the reactants, products and catalyst are molecularly dispersed in a single phase, usually the liquid phase. The third type is enzyme catalysis.

A challenging and timely problem is how to apply catalysis in air-pollution control. At the moment it appears that catalysis will be heavily involved in the treatment of exhaust gas from internal-combustion engines. This will represent the first instance of the direct application by the general public of prepared catalysts, as distinguish from natural catalysts such as yeast.

Selected from "Specialized Chemical English (unformal published) (the first volumn), by Cui Bo etc., 1994,70-74"

Exercise

1. Put the following into Chinese:

   completion   ammonia   hypothetical   oxidation
   chemisorb   susceptible   activity   calcination
   dehydrogenation   extrude   impregnate   olefin

2. Put the following into English:

   氯化钠   加速   二甲醚   还原
   选择性   氢化   烃   烃化
   半导体   异构化   气态的   聚合

Reading Material

Classification of Catalysts

From an accumulation of practical experience it is possible to narrow the range of solids that are likely to be catalysts for a type of chemical reaction. We discuss a few cases here and give information for more reactions in Table 16-1.

Table 16-1   Catalysts for Commercial Processes

Process	Typical catalysts	Poisons
Alkylation of hydrocarbons	$H_2SO_4(l)$, $HF(l)$, $AlCl_3+HCl$ H_3PO_4/kieselguhr	Substances which reduce acidty
Hydrocarbon cracking	Crystalline synthetic SiO_2-Al_2O_3(zeolites)	Nitrogen compounds, Metals(Ni.V.Cu) coke deposition
Chlorination of hydrocarbons	$CuCl_2$/Al_2O_3	Coke deposition
Dehydration	γ-Al_2O_3, SiO_2-Al_2O_3, WO_3	
Dehydrogenation	Cr_2O_3/Al_2O_3, Fe, Ni, Co, ZnO, Fe_2O_3	H_2O
Desulfurization of petroleum fractions	Sulfided Co-Mo/Al_2O_3	
Fischer-Tropsch process	Ni/kieselguhr, Fe+Fe_2O+Fe_3O_4	Sulfur, arsenic coke deposition
Hydrogen from naphtha, coal	Ni/refractory	sulfur, chlorine compounds
Hydrogenation	Ni/kieselguhr, NiO, Ni-Al(Raneykel) Pt/Al_2O_3, Pd/Al_2O_3, Ru/Al_2O_3	
Hydrocracking of coal, heavy oil	NiS, Co_2O_3-MoO_3/Al_2O_3 W_2O_3, $ZnCl_2$	
Isomerization	$AlCl_3+HCl$, Pt/Al_2O_3	
Oxidation inorganic	Pt, V_2O_5, Rh, $CuCl_2$ (HCl to Cl_2)	Arsenic, chlorine compounds
Oxidation organic (liquid phase)	$CuCl_2$(aq)+$PdCl_2$, Pd/Al_2O_3, Co+Cu acetates	
Oxidation organic (gas phase)	V_2O_5/Al_2O_3, Ag-AgO, CuO Bismuth molybdate	
Polymerization	$Al(C_2H_5)_3$, P_2O_5/kieselguhr, MoO_3-CoO/Al_2O_3, CrO_3/(SiO_2-Al_2O_3), $TiCl_3$-$Al(C_2H_5)_3$	H_2O, O_2, Surfur compounds CO, CO_2

Metals chemisorb oxygen and hydrogen and therefore are usually effective catalysts for oxidation-reduction and hydrogenation-dehydrogenation reactions. Thus platinum is a successful catalyst for the oxidation of SO_2 and Ni is used effectively for hydrogenation of hydrocarbons. The metals oxides, as semiconductors catalyze the same kinds of reactions, but often higher temperatures are required. Because of the relative strength of the chemisorption bond with which such gases as O_2 and CO are attached to metals, these gases are poisons when less susceptible to such poisoning. Oxides of the transition metals, such as MoO_3 and Cr_2O_3, are good catalysts for polymerization of olefins. Also aluminum alkyl-titanium chloride [for example, $Al(C_2H_5)_3^{3+}TiCl_4$] constitutes an excellent catalyst for producing isotactic polymers from olefins. Alumina and silica catalysts are widely used for alkylation, isomerization, polymerization, and particularly for cracking of hydrocarbons. In each case the mechanism presumably involves carbonium ions formed at the acid sites on the catalyst. While the emphasis here has been on solid catalysts, liquid and gaseous acids, particularly H_2SO_4 and HF, are well-known alkylation and isomerization catalysts.

Often catalysts are specific. An important example is the effectiveness of iron for producing hydrocarbons from hydrogen and carbon monoxide (the Fischer Tropsch synthesis). Dual-function catalysts for isomerization and reforming reactions consist of two active substances in close proximity to each other. For example, Ciapetta and Hunter found that a silica-alumina catalyst upon which nickel was dispersed was much more effective in isomerizing n-hexane than silica alumina alone. The explanation depends on the fact that olefins are more readily isomerized than paraffin hydrocarbons. Nickel presumably acts as a dehydrogenating agent, producing hexene, after which the silica alumina isomerizes the hexene to isohexene. Finally, the nickel is effective in hydrogenating hexene back to isohexane.

Condensed from material in "Catalytic Processes and Proven Catalysts" by Charles L.Thomas. Academic Press, New York, 1970.

Much has and is being written on solid catalysts, and helpful sources of summary information are available++.

Catalyst Preparation

Experimental methods and techniques for catalyst manufacture are particularly important because chemical composition is not enough by itself to determine activity. The physical properties of surface area, pore size, particle size, and particle structure also have an influence. These properties are determined to a large extent by the preparation procedure. To begin with, a distinction should be drawn between preparations in which the entire material constitutes the catalyst and those in which the active ingredient is dispersed on a support or carrier having a large surface area. The first kind of catalyst is usually made by precipitation, gel formation, or simple mixing of the components.

Precipitation provides a method of obtaining the solid material in a porous form. It consists of adding a precipitation agent to solutions of the desired components. Washing, drying, and ususlly calcination and activation (or pretreatment) are subsequent steps in the process. For example, a magnesium oxide catalyst can be prepared by precipitating the magnesium from nitrate solution by adding sodium carbonate. The precipitate of $MgCO_3$ is washed, dried, and calcined to obtain the oxide. Such variables as concentration of the aqueous solutions, temperature, and time of the drying and calcining steps may influence the surface area, pore structure, and intrinsic activity of the final product. This illustrates the difficulty in reproducing catalysts and indicates the necessity of carefully following impurities, which may act as poisons.

A special case of the precipitation method is the formation of a colloidal precipitate which gels. The steps in the process are essentially the same as for the usual precipitation procedure. Catalysts containing silica and alumina are especially suitable for preparation by gel formation, since their precipitates are of a colloidal nature. Detailed techniques for producing catalysts through gel formation or ordinary precipitation are given by Ciapetta and Plank.

In some instances a porous material can be obtained by mixing the components with water, milling to the desired grain size, drying, and calcining. Such materials must be ground and sieved to obtain the proper particle size. A mixed magnesium and calcium oxide catalyst can be prepared in this fashion. The carbonates are milled wet in a ball machine, extruded, dried, and reduced by heating in an oven.

Catalyst carriers provide a means of obtaining a large surface area with a small amount of active material. This is important when expensive agents such as platinum, palladium, ruthenium, and silver are used. Berkman etal. have treated the subject of carriers in some detail.

The steps in the preparation of a catalyst impregnated on a carrier may include (1) evacuating the carrier; (2) contacting the carrier with the impregnating solution; (3) removing the excess solution; (4) drying; (5) calcination and activation. For example, a nickel hydrogenation catalyst can be prepared on alumina by soaking the evacuated alumina particles with nickel nitrate solution, draining to remove the excess solution, and heating in an oven to decompose the nitrate to nickel oxide. The final step (activation), reduction of the oxide to metallic nickel is best carried out with the particles in place in the reactor by passing hydrogen through the equipment. Activation in site prevents contamination with air and other gases which might poison the reactive nickel. In this case no precipitation was required. This is a desirable method of preparation, since thorough impregnation of all the interior surface of the carrier particles is relatively simple. However, if the solution used to soak the carrier contains potential poisons such as chlorides or sulfates, it may be necessary to precipitate the required constituent and wash out the possible poison.

The nature of the support can affect catalyst activity and selectivity. This effect presumably arises because the support can influence the surface structure of the atoms of dispersed catalytic agent. For example, changing from a silica to alumina carrier may change the electronic structure of deposited platinum atoms. This question is related to the optimum amount of catalyst that should be deposited on a carrier. When only a small fraction of a monomolecular layer is added, increases in amount of catalyst should increase the rate. However, it may not be helpful to para hydrogen with a NiO catalyst deposited on alumina was found to be less for $5.0wt\%$ NiO than for $0.5wt\%$ NiO. The dispersion of the catalyst on the carrier may also be an important factor in such cases. The nickel atoms were deposited from a much more concentrated $NiNO_3$ solution to make the catalyst containing $5.0wt\%$ NiO. This may have led to larger clusters of nickel atoms. That is, many more nickel atoms were deposited on top of each other, so that the dispersion of nickel on the surface was less uniform than with the $0.5wt\%$ catalyst. It is interesting to note that a $5.0wt\%$ NiO catalyst prepared by 10 individual depositions of $0.5wt\%$ was much more active (by a factor of 11) than the $5.0wt\%$ added in a single treatment. The multiple deposition method presumably gave a much larger active nickel surface, because of better dispersion of the nickel atoms on the Al_2O_3 surface. Since the total amount of nickel was the same for the two preparations, one would conclude that the individual particles of nickel were smaller in the 10-application catalyst. These kinds of data indicate the importance of measuring surface areas for chemisorption of the reactants involved. A technique based on the chemisorption of H_2 and CO has been developed to study the effect of dispersion of a catalyst on its activity and the effect of interaction between catalyst and support on activity.

Selected from "Specialized Chemical English (unformal published) (volumn two), by Cui Bo etc., 1994, 104-112."

Lesson 17

Colloid

Colloid, a mixture in which one substance is divided into minute particles (called colloidal particles) and dispersed throughout a second substance. The mixture is also called a colloidal system, colloidal solution, or colloidal dispersion. Familiar colloids include fog, smoke, homogenized milk, and ruby-colored glass.

Colloids, Solutions, and Mixtures. The Scottish chemist Thomas Graham discover (1860) that certain substances (e.g., glue or gelatin) could be separated from certain other substances (e.g., sugar or salt) by dialysis. He gave the name colloid to substances that do not diffuse through a semipermeable membrane and the name crystalloid to those which do diffuse and which are therefore in true solution[1]. Colloidal particles are larger than molecules but too small to be observed directly with a microscope; however, their shape and size can be determined by electron microscopy. In a true solution the particles of dissolved substance are of molecular size and are thus smaller than colloidal particles; in a coarse mixture the particles are much larger than colloidal particles. Although there are no precise boundaries of size between the particles in mixtures, colloids, or solutions, colloidal particles are usually on the order of 10^{-7} to 10^{-5} cm in size.

Classification of Colloids. One way of classifying colloids is to group them according to the phase of the dispersed substance and of the medium of dispersion. A gas may be dispersed in a liquid to form a foam or in a solid to form a solid foam (e.g., styrofoam). A liquid may be dispersed in a gas to form an aerosol (e.g., fog or aerosol spray), in another liquid to form an emulsion (e.g., homogenized milk), or in a solid to form a gel (e.g., jellies or cheese). A solid may be dispersed in a gas to form a solid aerosol (e.g., dust or smoke in air), in a liquid to form a sol (e.g., ink or muddy water), or in a solid to form a solid sol (e.g., certain alloys).

A further distinction is often made in the case of adispersed solid. In some cases (e.g., a dispersion of sulfur in water), the colloidal particles have the same internal structure as a bulk of the solid. In other cases (e.g., a dispersion of soap in water), the particles are an aggregate of small molecules and do not correspond to any particular solid structure. In still other cases (e.g., a dispersion of a protein in water), the particles are actually very large single molecules. A different distinction, usually made when the dispersing medium is a liquid, is between lyophilic and lyophobic systems. The particles in a lyophilic system have a great affinity for the solvent, and are readily solvated (combined, chemically or physically, with the solvent) and dispersed, even at high concentrations[2]. In a lyophobic system the particles resist solvation and dispersion in the solvent, and the concentration of particles is usually relatively low.

Formation of Colloids. There are two basic methods of forming a colloid: reduction of larger particles to colloidal size, and condensation of smaller particles (e.g., molecules) into colloidal particles. Some substances are easily dispersed to form a colloid; this spontaneous dispersion is called peptization. A metal can be dispersed by evaporating it in an electric arc; if the electrodes are immersed in water, colloidal particles of the metal form as the metal vapor cools[3].A solid can be reduced to colloidal particles in a colloid mill, a mechanical device that uses a shearing force to break apart the larger particles. An emulsion is often prepared by homogenization, usually with the addition of an emulsifying agent. The above methods involve breaking down a larger substance into colloidal particles. Condensation of smaller particles to form a colloid usually involves chemical reactions—typically displacement, hydrolysis, or oxidation and reduction.

Properties of Colloids. One property of colloid systems that distinguishes them from true solutions is that colloidal particles scatter light. If a beam of light, such as that from a flashlight, passes through a colloid, the light is reflected by the colloidal particles and the path of the light can therefore be observed. When a beam of light passes through a true solution there is so little scattering of the light that the path of the light cannot be seen and small amount of scattered light cannot be detected except by very sensitive instruments. The scattering of light by colloids, known as the Tyndall effects, was first explained by the British physicist John Tyndall.

When an ultramicroscope is used to examine a colloid, the colloidal particles appear as tiny points of light in constant motion; this motion, called Brownian movement, helps keep the particles in suspension.

Absorption is another characteristic of colloids, since the finely divided colloidal particles have a large surface area exposed. The presence of colloidal particles has little effect on the colligative properties of a solution.

The particles of a colloid selectively absorb ions and acquire an electric charge. All of the particles of a given colloid take on the same charge and thus are repelled by one another. If an electric potential is applied to a colloid, the charged colloidal particles move toward the oppositely charged electrode; this migration is called electrophoresis. If the charge on the particles is naturalized, they may precipitate out of the suspension. A colloid may be precipitated by adding another colloid with oppositely charged particles; the particles are attracted to one another, coagulate, and precipitate out[4]. Addition of soluble ions may precipitate a colloid; the ions in seawater precipitate the colloidal silt dispersed in river water, forming a delta. Particles in a lyophobic system are readily coagulated and precipitate, and the system cannot easily be restored to its colloidal state. A lyophilic colloid does not readily precipitate and can usually be restored by the addition of solvent.

Thixotropy is a property exhibited by certain gels (semisolid, jelly like colloids). A thixotropic gel appears to be solid and maintains a shape of its own until it is subject to a shearing force or some other disturbance, such as shaking[5].It then acts as a sol (a semifluid colloid) and flows freely. Thixotropic behavior is reversible, and when allowed to stand undisturbed the sol slowly reverts to a gel. Common thixotropic gels include oil well drilling mud, certain paints and printing inks, and certain clay.

Stability of Colloidal Systems, Aggression, Coagulation, Flocculation. The terms stable and stability are used in rather special and often different senses in colloid science: the relationship between these usages and the formal thermodynamic usage is outlined below.

Thermodynamically stable or metastable means that the system is in a state of equilibrium corresponding to a local minimum of the appropriate thermodynamic potential for the specified constraints on the system (e.g., Gibbs free energy at constant temperature and pressure). Stability cannot be defined in an absolute sense, but if several states are in principle accessible to the system under given condition, that with the lowest potential is called the stable state, while the other states are described as metastable. Unstable states are not at a local minimum. Transitions between metastable and stable states occur at rates which depend on the magnitude of the appropriate activation energy barriers which separate them. Most colloidal systems are metastable or unstable with respect to the separate bulk phase, with the (possible) exception of lyophilic sols, gels and xerogels of macromolecules.

Colloidally stable means that the particles do not aggregate at a significant rate: the precise connotation depends on the type of aggregation under consideration. For examples, a concentrated paint is called stable by some people because oil and pigment do not separate out at a measurable rate, and unstable by others because the pigment particles aggregate into a continuous network.

An aggregate is, in general, a group of particles (which may be atoms or molecules) held together in any way: a colloidal particle itself (e.g., a micelle, see below) may be regarded as an aggregate. More specially, aggregate is used to describe the structure formed by the cohesion of colloidal particles.

Aggregation is the process or the result of the formation of aggregates.

When a sol is colloidally unstable the formation of aggregates is called coagulation or flocculation. These terms are often used interchangeably, but some authors prefer to introduce a distinction between coagulation, implying the formation of compact aggregates, leading to the macroscopic separation of a coagulum; and flocculation, implying the formation of a loose or open network which may or may not separate macroscopically. In many contexts the loose structure formed in this way is called a floc. While this distinction has certain advantages, in view of the more general (but not universal) acceptance of the equivalence of the words coagulation and flocculation, any author who wishes to make a distinction between them should state so clearly in his publication.

The reversal of coagulation or flocculation, for example, the dispersion of aggregates to form a colloidally stable suspension or emulsion, is called deflocculation (sometimes peptization).

The rate of aggregation is in general determined by the frequency of collisions and the probability of cohesion during collision. If the collisions are caused by Brownian motion, the process is called perikinetic aggregation; if by hydrodynamic motions (e.g., convection or sedimentation) one may speak of orthokinetic aggregation.

In hydrophobic sols, coagulation can be brought about by changing the electrolyte concentration to the critical coagulation concentration (CCC). As the value of the critical coagulation concentration depends to some extent on the experimental circumstances (method of mixing, time between mixing and determining the state of coagulation, criterion for measuring the degree of coagulation, etc.) these should be clearly stated[6].

The generalization that the critical coagulation concentration for a typical lyophobic sol is extremely sensitive to the valence of the counterions (high valence gives a low critical coagulation concentration) is called the Schulze-Hardy rule.

Addition of small amounts of a hydrophilic colloid to a hydrophobic sol may make the latter more sensitive to flocculation by electrolyte. This phenomenon is called sensitization. Higher concentrations of the same hydrophilic colloid usually protect the hydrophobic sol from flocculation. This phenomenon is called protective action. Colloidally stable mixtures of a lyophobic and lyophilic colloid are called protected lyophobic colloids; although they may be thermodynamically unstable with respect to macroscopic phase separation, they have many properties in common with lyophilic colloids.

Sedimentation is the settling of suspended particles under the action of gravity or a centrifugal field. If the concentration of particles is high and interparticle forces are strong enough, the process of sedimentation may be better described as compaction of the particle structure with pressing out of the liquid. This particular kind of settling is also called subsidence.

Sediment is the highly concentrated suspension which may be formed by the sedimentation of a dilute suspension.

Coalescence is the disappearance of the boundary between two particles (usually droplets or bubbles) in contact, or between one of these and a bulk phase followed by changes of shape leading to a reduction of the total surface area[7].The flocculation of an emulsion, viz. the formation of aggregates, may be followed by coalescence. If coalescence is executive it leads to the formation of a macrophase and the emulsion is said to break.

The breaking of a foam involves the coalescence of gas bubbles. Coalescence of solid particles is called sintering.

Creaming is the macroscopic separation of a dilute emulsion into a highly concentrated emulsion, in which interglobular contact is important, and a continuous phase under the action of gravity or a centrifugal field. This separation usually occurs upward, but the term may still be applied if the relative densities of the dispersed and continuous phases are such that the concentrated emulsion settles downward. Some authors, however, also use creaming as the opposite of sedimentation even when the particles are not emulsion droplets.

Cream is the highly concentrated emulsion formed by creaming of a dilute emulsion. The droplets in the cream may be colloidally stable or flocculated, but they should not have coalesced.

As a rule all colloidal system, initially of uniform concentration, establish, when subjected to the action of gravity or a centrifugal field, a concentration gradient as a result of sedimentation or creaming; but if the system is colloidally stable the particles in the sediment or cream do not aggregate and can be redispersed by the application of forces of the same magnitude as those which caused sedimentation or creaming.

The loss of the stability of a lyophilic sol (equivalent to a decrease in the solubility of the lyophilic colloid) quite often results in a separation of the system into two liquid phases. The separation into two liquid phases in colloidal systems is called coacervation. It occurs also, though rarely, in hydrophobic sols. The phase more concentrated in colloid component is the coacervate, and the other phase is the equilibrium solution.

Selected from "English for Chemistry and Chemical Engineering, by Zhang Yuping etc., Chemical Industry Press, 2007,213-218"

New Words

1. colloid ['kɔlɔid] *n.* 胶体；*a.* 胶体的
2. gelatin ['dʒelətin] *n.* 凝胶，白明胶
3. dialysis [dai'ælisis] *n.* 渗透，透析
4. semipermeable ['semi'pɜːmiəbl] *a.* 半透性的
5. crystalloid ['kristə,lɔid] *a.* 结晶的；*n.* 晶体
6. microscope [,maikrəskəup] *n.* 显微镜
7. coarse [kɔːs] *a.* 粗糙的，粗鄙的
8. styrofoam ['stairəfəum] *n.* 泡沫聚苯乙烯
9. aerosol ['ɛərəsɔl] *n.* 气溶胶，悬浮微粒
10. emulsion [i'mʌlʃən] *n.* 乳状液
11. distinction [dis'tiŋkʃən] *n.* 区别，差别，特性
12. lyophilic [,laiə'filik] *a.* 亲液的
13. lyophobic [,laiə'fɔbik] *a.* 疏液的，憎液的
14. peptization [,peptai'zeiʃən] *n.* 胶溶作

用，解胶（作用），塑解
15. evaporating [iˈvæpəˌreitiŋ] *a.* 蒸发用的，蒸发作用的
16. condensation [ˌkɔndenˈseiʃən] *n.* 凝结，凝固，浓缩
17. scatter [ˈskætə] *v.* 散射，分散，散开
18. colligative [ˈkɔligətiv] *a.* 依数的
19. electrophoresis [iˌlektrəfəˈriːsis] *n.* 电泳
20. precipitate [priˈsipiteit] *n.* 沉淀物；*v.* 沉淀
21. thixotropy [θikˈsɔtrəpi] *n.* 触变性
22. coagulation [kəuˌægjuˈleiʃən] *n.* 凝结，絮（胶）凝，聚沉
23. flocculation [ˌflɔkjuˈleiʃən] *n.* 絮凝，絮接

产物
24. metastable [ˌmetəˈsteibl] *a.* 亚稳的
25. xerogel [ˈziərədʒel] *n.* 干凝胶
26. perikinetic [ˌperikaiˈnetik] *a.* 异向的，与布朗运动有关的
27. orthokinetic [ˌɔːθəkaiˈnetik] *a.* 同向移动的
28. centrifugal [senˈtrifjugəl] *a.* 离心的
29. subsidence [ˈsʌbsidəns] *n.* 沉淀，陷没，下沉
30. coalescence [ˌkəuəˈlesəns] *n.* 聚结
31. connotation [ˌkɔnəˈteiʃən] *n.* 涵义，含义

Phrases

be divided into… 被分为……   in view of… 鉴于……，考虑到……
colloid mill   胶体磨

Affixes

-scope 构成名词，表示"镜、显示器"，如：microscope, telescope
-fy 构成动词，表示"（使）成为"，"产生"，"使感动" 如：beautify, classify, intensify
peri- 表示"周围，近"之意，如：perigon, perimetry
ortho- 表示"直，正直"之意，如：orthocenter, orthodox, orthogonal, orthogenics

Notes

1. He gave the name colloid to substances that do not diffuse through a semipermeable membrane and the name crystalloid to those which do diffuse and which are therefore in true solution. that do…membrane 修饰 substances; which…which…用来修饰 those（substances）。参考译文：Thomas Graham 将不能扩散通过半透膜的物质叫做胶体，将能通过半透膜，也就是存在于真溶液中的物质叫做晶体。

2. The particles in a lyophilic system have a great affinity for the solvent, and are readily solvated (combined, chemically or physically, with the solvent) and dispersed, even at high concentrations. 参考译文：亲溶液胶中的胶体粒子与溶剂有较强的亲和力，易于溶剂化（即与溶剂通过物理或化学作用结合）和扩散，即使溶胶的浓度很高。

3. A metal can be dispersed by evaporating it in an electric arc; if the electrodes are immersed in waters, colloidal particles of the metal form as the metal vapor cools. as 表示"当……的时候"。参考译文：金属可以通过电弧使之蒸发而分散，如果电极浸入水中，当金属蒸气冷却时，就会形成金属胶体颗粒。

4. A colloid may be precipitated by adding another colloid with oppositely charged particle; the particles are attracted to one another, coagulate, and precipitate out. 参考译文：通过在溶胶中加入与其胶粒电性相反的溶胶使之沉淀；这些胶粒因为电性不同而相互吸引、凝结，进而从溶胶

中沉淀出来。

5. A thixotropic gel appears to be solid and maintains a shape of its own until it is subject to a shearing force or some other disturbance, such as shaking. subject to 意思是"使服从，使遭受"。句中的 it 都指 gel。参考译文：具有触变性的凝胶外表保持固体的形状直到它遭受到剪切力或摇动等其他干扰。

6. As the value of the critical coagulation concentration depends to some extent on the experimental circumstances(method of mixing, time between mixing and determining the state of coagulation, criterion for measuring the degree of coagulation, etc.) these should be clearly stated. as 引导的分句作原因状语，these 指代括号中的条件。参考译文：由于临界聚沉浓度在一定程度上取决于实验环境（混合的方法，从混合到确认到达凝结状态的时间以及凝结程度的测量标准等），这些实验条件都必须明确规定。

7. Coalescence is the disappearance of the boundary between two particles (usually droplets or bubbles) in contact, or between one of these and a bulk phase followed by changes of shape leading to a reduction of the total surface area. 参考译文：聚结指两个相连的粒子（通常是液滴或者泡沫）之间的边界消失，或者其中的一种粒子与凝聚相之间的界面消失，从而引起该物质的形状改变导致总表面积减少。

Exercise

1. Put the following into Chinese:

colloid	condensation	centrifugal	sulfate
crystalloid	scatter	metastable	granular
coarse	coagulation	contaminant	anionic

2. Put the following into English：

渗析	乳状液	电泳	阳离子的
半透性的	亲液的	沉淀	氢氧化物
显微镜	疏液的	亚铁的	腐蚀性

Reading Material

Coagulation and Flocculation

Removal of turbidity by coagulation depends on the nature and concentration of the colloidal contaminants; type and dosage of chemical coagulant; use of coagulant aids; and chemical characteristics of the water, such as pH, temperature, and ionic character. Because of the complex nature of coagulation reactions, chemical treatment of water supplies is based primarily on empirical data derived from laboratory and field studies. However, in recent years a considerable amount of research has been directed toward gaining a better understanding of coagulation mechanisms.

In destabilizing colloids two basic mechanisms have been described as helping form sufficiently, large aggregates to facilitate settling from suspension. The first, referred to as coagulation, reduces the net electrical repulsive forces at particle surfaces by electrolytes in solution. The second mechanism, known as flocculation, is aggregation by, chemical bridging between particles. In water treatment particle, chemical coagulation and flocculation are also considered to depend on physical processes. Choice of coagulation dosage, pH, and coagulant aids are related to the mixing process promoting aggregation of the destabilized colloids. The efficiency, of the coagulation-flocculation system depends on subsequent settling and filtration. Trace quantities of impurities frequently present in natural waters (e.g. color, and silica) can have significant effects on both the chemical and physical properties of flocs formed during coagulation-flocculation and thereby alter their settling and filtering characteristics.

Traditionally, sanitary engineers have not restricted the use of terms coagulation and flocculation to describing chemical mechanisms only. Common use of these terms refers to both chemical and physical processes in treatment, possibly because the complex reactions that take place in chemical coagulation-flocculation are only partially understood. More importance is the fact that engineers tends to tie coagulation to the operational units (mixing devices and flocculation chambers) used in chemical treatment. The physical processes of mixing and flocculation are described in following sections. Coagulation concerns the series of chemical and mechanical operations by which coagulants are applied and made effective. These operations are customarily considered to comprise two distinct phases: (1) mixing, wherein the dissolved coagulant is rapidly dispersed throughout the water being treated, usually by violent agitation, and (2) flocculation, involving agitation of the water at lower velocities for a much longer period, during which very small particles grow and agglomerate into well-defined flocs of sufficient size to settle readily.

1. Coagulants

The most widely used coagulants for water and wastewater treatment are aluminum and iron salts. The common metal salt is aluminum sulfate (filter alum), which is a good coagulant for water containing appreciable organic matter. The material is commonly shipped and fed in a dry granular form, although it is available as a powder or liquid alum syrup. Aluminum sulfate reacts with natural alkalinity in water to form aluminum hydroxide floc. If water does not contain sufficient alkalinity to react with the alum, lime or soda ash is fed to provide the necessary alkalinity. An advantage of using sodium carbonate (soda ash) is that unlike lime it does not increase water hardness, only corrosiveness.

Iron coagulants operate over a wider pH range and are generally more effective in removing color from water; however, they are usually more costly. They include ferrous sulfate (copperas), ferric sulfate, and ferric chloride. Ferrous sulfate reacts with natural alkalinity, but the response is much slower than that between alum and natural alkalinity. Lime is generally added to raise the pH to the point where ferrous ions are precipitated as ferric hydroxide by the caustic alkalinity. Advantages of the ferric coagulants are that (1) coagulation is possible over a wide pH range, generally pH 4~9 for most waters; (2) the precipitate produced is a heavy quick settling floc; and (3) they are more effective in the removal of color, taste, and odor compounds. Ferric sulfate is available in crystalline form and may be fed using dry or liquid feeders. Although ferric chloride can be used in water treatment, its most frequent application is in wastewater treatment (e. g. as a waste sludge conditioning chemical m combination with lime prior to mechanical dewatering).

Cationic polyelectrolytes serve as prime coagulants, but their typical application is as a coagulant aid. Conventional aids adopted for metal coagulants involve lime, soda ash, activated silica and polyelectrolytes. The final choice of coagulant and chemical aids for particular water is based on coagulation control tests, past experience in treatment of water of similar quality, and the overall economics involved.

2. Coagulant Aids—Polyelectrolytes

A threefold classification of polyelectrolytes is based on their ionic character; negatively charged compounds are called anionic polyelectrolytes. Positively charged are cationic polymers, and compounds with both positive and negative charges referred to as polyampholytes. Commercially, available polyelectrolytes are manufactured by a number of companies under a variety of trade names.

Polyelectrolytes are effective as coagulant aids; however, the type of polyelectrolyte, dosage, and point of addition must be determined for each water. As coagulant aids, they increase the rate and degree of flocculation (aggregation) by adsorption, charge neutralization, and interparticle bridging. The latter appears to be the main mechanism by which polyelectrolytes aid coagulation.

3. Acids and Alkalies

Acids and alkalies are added to water to adjust pH for optimum coagulation. Typical acids used to lower the pH are sulfuric and phosphoric. Alkalies used to raise the pH are lime, sodium hydroxide, and soda ash. Hydrated lime, with about 70% available CaO, is suitable for dry feeding but costs more than quicklime, which is 90% CaO. The latter must be slaked (combined with water) and fed as a lime slurry. Soda ash is 98% sodium carbonate and can be fed dry but is more expensive than lime. Sodium hydroxide is purchased and fed as a concentrated solution.

Selected from "Specialized English for Applied Chemistry, by Zhu Hongjun ect., Chemical Industry Press, 2005, 238-240"

New words

1. contaminant [kən'tæminənt] n. 污染物
2. ferrous ['ferəs] a. (二价) 亚铁的
3. coagulant [kəu'ægjulənt] n. 混凝剂
4. ferric ['ferik] a. (三价) 铁的
5. destabilize [diː'steibilaiz] vt. 使脱稳
6. feeder ['fiːdə] n. 给水管，进料器
7. aggregate ['ægrigeit] n. 集合（体），聚集（体）
8. cationic a. 阳离子的
9. electrolyte [i'lektrəulait] n. 电解质
10. polyelectrolyte [ˌpɔliiˈlektrəuˌlait] n. 聚合电解质
11. aggregation [ægri'geiʃən] n. 聚合（作用），聚合体
12. anionic [ænai'ɔnik] a. 阴离子的
13. dosage ['dəusidʒ] n. 用（剂）量
14. threefold ad. 三倍（于），增量三倍
15. silica ['silikə] n. 二氧化硅
16. polyampholyte [ˌpɔli'æmfəˌlait] n. 聚两性电解质
17. disperse [dis'pəːs] v. 分散，弥散
18. sulfuric [sʌl'fjuərik] a. 硫的，硫黄的
19. sulfate [ˌsʌlfeit] n. 硫酸盐
20. phosphoric [fɔs'fɔrik] a. 磷的，含(五价)磷的
21. alum [ə'lʌm] n. (明，白)矾，硫酸铝
22. quicklime ['kwiklaim] n. 生石灰，氧化钙
23. carbonate ['kɑːbəneit] n. 碳酸盐
24. hydrate ['haidreit] v. (使)水合(化)(使成水合物)
25. granular ['grænjulə] a. 颗粒状的
26. hydroxide n. 氢氧化物
27. slake v. 消解，潮（水）解
28. corrosiveness n. 腐蚀性，侵蚀作用
29. be described as 说成是，被称作
30. copperas ['kɔpərəs] n. 绿矾，硫酸亚铁
31. restrict to 把……局限于……（范围）

Lesson 18

Polymers and Polymerization Techniques

Polymers are all around us. They are the main components of food (starch, protein), clothes (silk, cotton, polyester, nylon) dwellings (wood-cellulose, paints) and also our bodies (nucleic acids, polysaccharides, proteins). No distinction is made between biopolymers and synthetic polymers. Indeed many of the early synthetic polymers were based upon naturally occurring polymers, e.g. celluloid (cellulose nitrate), vulcanization of rubber, rayon (cellulose acetate).

Polymers are constructed form monomer units, connected by covalent bonds. The definition of a polymer is:

"a substance, —R—R—R—R—or, in general $-[R]_n$, where R is bifunctional entity (or bivalent radical) which is not capable of a separate existence."

Where n is the degree of polymerization, DP_n. This definition excludes simple organic and inorganic compounds, e.g. CH_4, NaCl, and also excludes materials like diamond, silica and metals which appear to have the properties of polymers, but are capable of being vaporized into monomer units.

The molecular weight (MW) (strictly relative molecular mass) can be obtained from the MW of the monomer (or repeat unit) multiplied by n. Thus the MW of CH_4 or NaCl is 18 or 58.5 respectively, whereas the MW of a polymer can be >100. When the value of n is small, say 2~20, the substances are called oligomers[1], often these oligomers are capable of further polymerizations and then are referred to as macromers.

By definition, 1 mole of a polymer contains 6×10^{27} polymer molecules and therefore 1g mole=MW of the polymer in grams, which, in theory, can be $>10^6$g. However, by convention, 1g mole usually refers to the MW of the repeat units; thus 1g mole of polyethylene $-(CH_2)_n-$ is taken as 14g (the end groups, being negligible, are ignore).

A polymer with a MW of 10^7, if fully extended, should have a length of ~1mm and a diameter of ~0.5nm. This is equivalent in size to uncooked spaghetti ~2km in length. However, in reality, in bulk polymers the chain is never fully extended—a random coil configuration is adopted sweeping out a space of diameter ~200nm. It therefore has the appearance of cooked spaghetti or worms (or more correctly, worms of different length). The movements of these polymer chains are determined by several factors, such as:

(i) Temperature

(ii) Chemical make-up of the backbone —C—C—C— chain; whether the chain is flexible (aliphatic structure) or rigid (aromatic)

(iii) The presence or absence of side-chains on the backbone

(iv) The inter-polymer chain attraction (weak-dipole/dipole, H-bonding —or strong covalent bonds, cross-linking)

(v) The *MW* and molecular weight distribution (*MWD*) of the polymer.

Nearly all of the properties of polymers can be predicted if the above factors are known, e.g. whether the polymer is amorphous or partially crystalline; the melting temperature of the crystalline phase (T_m) (actually it is more of a softening temperature over several degrees); is the polymer brittle or tough; its rigidity or stiffness (called modulus), whether the polymer dissolves in solvents, etc.

Polymers are really effect chemicals in that they are used as materials, e.g. plastics, fibers, films, elastomers, adhesives, paints, etc., with each application requiring different polymer properties. Many of the initial uses of plastics were inappropriate, which led to the belief that plastics were "cheap and nasty". However, recent legislation on product liability and a better understanding of the advantages and disadvantages of plastics have changed this position.

Economics, that is the cost of making and fabricating the polymer is of prime importance. This has led to a rough grouping of polymers into commodity polymers, engineering polymers, and advanced polymeric materials.

1. Commodity Polymers

Examples of these are:

Polyethylene $\begin{cases} \text{Low density polyethylene (LDPE)} \\ \text{High density polyethylene (HDPE)} \\ \text{Linear low density polyethylene (LLDPE)} \end{cases}$

Polypropylene (PP)

Poly vinyl chloride (PVC)

Polystyrene (PS)

Each of these is prepared on the 10 million tonnes /year scale. The price is < $ 1500/tonne.

2. Engineering Polymers

The materials have enjoyed the highest percentages growth of any polymers in the last ten years and are principally used as replacements for metals for moderate temperature (<150℃) and environmental conditions or they may have outstanding chemical inertness and/or special properties, e.g. low friction polytetrafluoroethylene(PTFE). These engineering polymers include

Acetal (or polyoxymethylene, POM)

Nylons (polyamides)

Polyethylene or polybutylene terephthalate (PET or PBT)

Polycarbonate (of bisphenol A) (PC)

Polyphenylene oxide (PPO) (usually blended with styrene)

The price are ($ 3000～ $ 15000) /tonne.

3. Advanced Polymeric Materials

These have very good temperature stability (many hours/days at 250～300℃) and when reinforced with fibers (e.g., glass, carbon or Aramid fibres)[2], i.e. composites, they are stronger than most metals on a weight/weight basis. They are usually only used sparingly, often in critical parts of a structure. Their price can be as high as $ 150, 000 /tonne.

4. Making of Polymers

Approximately 100 million tonnes of polymers are made annually, in plants ranging from 240,000 tonnes/year continuous single-stream polypropylene plants to a single batch preparation of a few kilograms of advanced performance composites. The highest tonnage polymers are LDPE, HDPE, LLDPE, PP, PVC, and PS.

The most important parameters in making polymers are quality control and reproducibility. They are different from simple organic compounds such as acetone, when often a simple distillation gives the desired purity. There are many different grades of the "same" polymer, depending on the final application, e.g., different *MW*, *MWD*, extent of branching, cross-linking, etc., and these variations are multiplied when copolymers (random, alternating and block) are considered. Many of the properties are fixed during polymerization and cannot be altered by post-treatment. Blending is sometimes carried out to obtain desired properties or just to up-grade production polymer that may be lightly off-specification.

A polymerization process consists of three stages:

(1) *Monomer preparation*. This is not discussed here, other than to emphasize that the purity of the monomer is paramount. Thus monofunctional impurities in step-growth, and radical scavengers, chain-transfer impurities and catalyst poisons (e.g., water in ionic propagation) in chain polymerizations are all significant.

(2) *Polymerization*. As stated above, uniformity of polymer properties is absolutely necessary, this not only includes *MW*, etc., but other factors such as color, shape of polymer particle (if the polymer is not palletized or granulated), catalyst residue, odor (especially when used for food application), etc.

The polymerization operation has to cope with the following parameters.

(i) Homogeneous or heterogeneous reactions.

(ii) In homogeneous systems, large increases in viscosity, affecting kinetics of polymerization, heat transfer and efficiency of mixing.

(iii) Most chain-growth polymerizations are exothermic, giving out several hundred kilowatts per ton of polymer produced; this heat has to be removed, since most polymerization are performed at constant temperature (isothermal). Heat removal is achieved by heat transfer through reactor walls, reactor design (long tubular reactors with high surface area/volume ratio), added cold reagents (monomer and/or solvent) or by using the latent heat of evaporation of monomer or solvents.

(iv) Control of *MW* and *MWD*, branching and cross-linking in the polymer. The polymerization process affects these whether batch, semi-batch (polymerizing reaction mixture is fed through to further reactors) or continuous. The residence time[3] of polymerization, whether narrow or broad, also determines *MW* and *MWD*.

As a general rule, once a process has been perfected it is not altered, unless economics dictate a change—for example, gas-phase polymerization performed in the absence of a solvent, thus eliminating solvent purification, recovery, fire hazard, etc.

(3) *Polymer recovery*. Unless the polymerization takes place in bulk, separation from the solvent has to be carried out. The conventional methods for recovering chemicals e.g., crystallization, distillation, adsorption, etc. are not normally to be used because polymers possess properties such as high viscosity and low solubility in solvents, and are sticky and non-volatile. Nevertheless, precipitation by using a non-solvent followed by centrifuging, or by coagulation of an emulsion or latex and removal of the solvent by steam-stripping (to keep the temperature down and prevent decomposition) can be used. Devolatilisation of the solvent or unused monomer from the polymer can be performed during pelletization in an extruder.

5. Polymerization Techniques

Most polymerizations are performed in the liquid phase using either a batch or a continuous process. The continuous method is preferred because it lends itself to smoother operation leading to a more uniform product, because of modern on-line analysis techniques. It also has lower operating costs. However, continuous processes have difficulties. The residence time of the polymer in the reactor will be variable, unless plug-flow is adopted using tube reactors. This may result in:

(ⅰ) Catalyst residues in the polymers, e.g., peroxides which may degrade the polymer during granulation or processing.

(ⅱ) The polymer may have a broad *MWD* with some decomposition, e.g., cross-linking (for long residence times).

(ⅲ) The polymer may adhere to the reactor walls, requiring shutdown for cleaning.

(ⅳ) Repeated changes in polymer grade may be required, during change-over the polymer will be a mixture making the polymer unsuitable for use.

There are five general methods of polymerization:
(ⅰ) Bulk (or mass)
(ⅱ) Solution
(ⅲ) Slurry (or precipitation)
(ⅳ) Suspension (or dispersion)
(ⅴ) Emulsion

Further lesser-used methods include:
(ⅵ) Interfacial
(ⅶ) Reaction injection moulding (RIM)[4]
(ⅷ) Reactive processing of molten polymers.

Selected from "An Introduction to Industrial Chemistry, 2nd Edition, C. A. Heaton, Blackie & Son Ltd., 1997" and "The Fontana History of Chemistry, William H. Brock, Fontana Press, 1992"

New Words

1. nucleic acid   核酸
2. biopolymer   *n.* 生物聚合物
3. celluloid ['seljuloid]   *n.* 赛璐珞，硝酸纤维素
4. vulcanization ['vʌlkənai'zeiʃən]   *n.* 硫化，硬化
5. rayon ['reiɔn]   *n.* 人造丝，人造纤维
6. bifunctional   *a.* 双官能团的
7. entity ['entiti]   *n.* 存在，实体，统一体
8. degree of polymerization   聚合度
9. diamond ['daiəmənd]   *n.* 金刚石，钻石
10. oligomer   *n.* 低聚物
11. macromer   *n.* 高聚物
12. spaghetti [spə'geti]   *n.* 通心粉
13. backbone ['bækbəun]   *n.* 构架，骨干，主要成分
14. dipole ['daipoul]   *n.* 偶极（子）
15. H-bond   *n.* 氢键
16. cross-link ['krɔːsliŋk]   *v.*（聚合物）交联，横向耦合
17. amorphous [ə'mɔːfəs]   *a.* 无定形的，非晶体的
18. brittle ['britl]   *a.* 脆的，易碎的
19. stiffness ['stifnis]   *n.* 刚性度，韧性
20. modulus ['mɔdjuləs]   *n.* 模数，系数，指数
21. liability [laiə'biliti]   *n.* 责任，义务
22. fabricate ['fæbrikeit]   *vt.* 制造，生产，制备；装配，安装，组合
23. linear ['liniə]   *a.*（直）线的，直线形的，线性的，线性化的
24. poly vinyl chloride   聚氯乙烯
25. polystyrene [pɔli'staiəriːn]   聚苯乙烯
26. acetal ['æsitæl]   *n.* 乙缩醛，乙醛缩二乙醇

27. polyoxymethylene *n.* 聚甲醛，聚氧化亚甲基
28. polybutylene terephthalate 聚丁烯对苯二酸酯
29. polycarbonate *n.* 聚碳酸酯
30. bisphenol A 双酚 A
31. polyphenylene oxide 聚苯氧化物
32. composite ['kɔmpəzit] *n.* 复合材料，合成[复合，组合]物
33. reproducibility *n.* 再生性，还原性，重复性
34. copolymer [kəu'pɔlimə] *n.* 共聚物
35. randam ['rændəm] *a.* 随机的，无规则的，偶然的
36. scavenger ['skævindʒə] *n.* 清除剂，净化剂
37. propagation [prɔpə'geiʃən] *n.* 增长，繁殖，传播，波及
38. pelletize ['pelitaiz] *v.* 造粒，做成丸（球，片）状
39. granulate ['grænjuleit] *vt.* 使成颗粒，使成粒状
40. heherogeneous ['hetərəu'dʒi:njəs] *a.* 多相的，非均匀的
41. kilowatt ['kiləwɔt] *n.* 千瓦（特）
42. isothermal [aisəu'θə:məl] *a.*; *n.* 等温（的），等温线（的）
43. latent heat 潜热
44. residence time 停留时间
45. adsorption [æd'sɔpʃən] *n.* 吸附（作用）
46. centrifuge ['sentrifjudʒ] *n.*; *v.* 离心，离心机，离心器
47. coagulation [kəuægju'leiʃən] *n.* 凝结（聚，固），胶凝，絮凝
48. latex ['leiteks] *n.* 橡胶，乳状物，（天然橡胶，人造橡胶）乳液
49. devolatilization *n.* 脱（去）挥发分（作用）
50. extruder [eks'tru:də] *n.* 挤压机，（螺旋）压出机
51. peroxide [pə'rɔksaid] *n.* 过氧化物
52. bulk polymerization 本体聚合
53. solution polymerization 溶液聚合
54. slurry polymerization 淤浆聚合
55. suspension polymerization 悬浮聚合
56. emulsion polymerization 乳液聚合
57. interficial polymerization 界面聚合
58. reaction injection moulding 反应注射成型
59. reactive processing of molten polymers 熔融聚合物的反应加工

Notes

1. oligomer:低聚物。单体聚合产生分子量较低的低聚物，称为低聚反应（oligomerization），产物称为低聚物。

2. aramid fibers:芳香族聚酰胺类纤维。由含有重复出现的酰胺基团（CO—NH—）直接连接到两个芳香环的线性聚合物所纺制的纤维（Aramid 是由 aromatic amide 缩写而成）。这种纤维具有特别高的强度，常用于复合材料。

3. residence time（in reactor）：停留时间，指物料在反应器中的停留时间。在连续操作设备中，由于物料的逆向混合（返混，backmixing），在同一时刻进入设备的物料可能分别取不同的流动路径，在设备内的停留时间也不相同，从而形成一定的分布，称为停留时间分布（residence time distribution）。

4. reaction injection moulding：反应注射成型，简称 RIM，成型过程中有化学反应的一种注射成型方法。这种方法所用原料不是聚合物，而是将两种或两种以上的液体单体或预聚物，以一定比例分别加到混合头中，在加压下混合均匀，立即注射到闭合模具中，在模具中聚合固化，定型成制品。

Exercises

1. Put the following into Chinese:
   oligomer   macromer   copolymer   propagation

uvlcanization	stiffness	fabricate	linear
reproducibility	residence time	coagulation	foresight
coordination	stereochemical	plug flow	injection-moulding
antiknock	alkylation	finishing	desalt
differentiate	diesel oil	lubricating oil	precursor
stripper	carbonium	radical	predominate

2. Put following into English:

官能团	单体	构架	模数
复合材料	非均相的	潜热	显热
热固性的	热塑性	无定形的	交联
随机的	等温的	吸附	离心
管式的	加氢裂解	异构化	组成
热解	腐蚀	残余物	液化石油气

Reading Material

Petroleum Processing

Petroleum, the product of natural changes in organic materials over millennia, has accumulated beneath the earth's surface in almost unbelievable quantities and has been discovered by humans and used to meet our varied fuel wants. Because it is a mixture of thousands of organic substances, it has proved adaptable to our changing needs. It has been adapted, through changing patterns of processing or refining, to the manufacture of a variety of fuels and through chemical changes to the manufacture of a host of pure chemical substances, the petrochemicals.

Modern units operate continuously. First a tubular heater supplies hot oil to an efficient distillation column which separates the material by boiling points into products similar to those obtained with the batch still, but more cleanly separated; then later units convert the less salable parts of the crude (the so-called bottom half of the barrel) into desired salable products. The processes used include various cracking units (which make small molecules from large ones), polymerization, reforming, hydrocracking, hydrotreating, isomerization, severe processing known as coking, and literally dozens of other processes designed to alter boiling point and molecular geometry.

1. Constituents of Petroleum

Crude petroleum is made up of thousands of different chemical substances including gases, liquids, and solids and ranging form methane to asphalt. Most constituents are hydrocarbons, but there are significant amounts of compounds containing nitrogen (0 to 0.5%), sulfur (0 to 6%), and oxygen (0 to 3.5%). No one constituent exists in large quantity in any crude.

Aliphatics, or Open Chain Hydrocarbons

n-Paraffin Series of n-Alkanes, C_nH_{2n+2}. This series comprises a larger fraction of most crudes than any other. Most straight-run (i.e., distilled directly from the crude) gasolines are predominantly n-paraffins. These materials have poor antiknock properties.

Iso-Paraffin Series or Iso-Alkanes, C_nH_{2n+2}. These branched chain materials perform better in internal-combustion engines than n-paraffins and hence are considered more desirable. They may be formed by catalytic reforming, alkylation, polymerization, or isomerization. Only small amounts exist in crudes.

Olefin, or Alkene Series, C_nH_{2n}. This series is generally absent in crudes, but refining processes such as cracking (making smaller molecules from large ones) produce them. These relatively unstable molecules improve the antiknock quality of gasoline, although not as effectively as iso-paraffins. On storage they polymerize and oxidized, which is undesirable. This very tendency to react, however, makes them useful for forming other compounds, petrochemicals, by additional chemical reactions. Ethylene, propylene, and butylene (also called ethene, propene, and butene) are examples. Cracked gasolines contain many higher members of the series.

Ring Compounds

Naphthene Series or Cycloalkanes, C_nH_{2n}. This series, not to be confused with naphthalene, has the same chemical formula as the olefins, but lacks their instability and reactivity because the molecular configuration permits them to be saturated and unreactive like the alkanes, these compounds are the second most abundant series of compounds in most crudes. The lower members of their group are good fuels; higher molecular weight ones are predominant in gas oil and lubricating oils separated from all types of crudes.

Aromatic, or Benzenoid Series, C_nH_{2n-6}. Only small amounts of this series occur in most common crudes, but they are very desirable in gasoline since they have high antiknock value, good storage stability, and many uses besides fuels. Many aromatics are formed by refining processes. Examples are: benzene, toluene, ehylbenzene, and xylene.

Lesser Components. Sulfur has always been an undesirable constituent of petroleum. The strong, objectionable odor of its compounds originally brought about efforts to eliminate them from gasoline and kerosene fraction. Chemical reactions were at first directed at destroying the odor. Later it was found that sulfur compounds had other undesirable effects (corrosion, reducing the effect of tetraethyl lead as an antiknock agent, air pollution). At present, wherever possible, the sulfur compounds are being removed and frequently the sulfur thus removed is recovered as elemental sulfur. Nitrogen compounds cause fewer problems than sulfur compounds, are less objectionable, and are generally ignored.

With the general adoption of catalytic cracking and finishing processes, it was discovered that the occurrence of metals present only in traces (Fe, Mo, Na, Ni, V, etc.) was troublesome as they are strong catalyst poisons. Now methods to remove these substances are being perfected. Salt has been a major problem for many years. It is practically always present in raw crude, usually as an emulsion, and must be removed to present corrosion. It breaks down heating in the presence of hydrocarbons to produce hydrochloric acid. Mechanical or electrical desalting is preliminary to most crude-processing steps.

Petroleum crudes vary widely, each kind requiring different refining procedures. The terms paraffin base, asphalt (naphthene), and mixed base are often applied to differentiate crudes on the basis of the residues produced after simple distillation.

Pure chemical compounds are not regularly separated by refining processes. Some of the simpler, low molecular weight ones are isolated for processing into petrochemicals. Most petroleum products are mixtures separated on the basis of boiling point ranges and identified by the ultimate uses to which they are well adapted.

Common refinery fractions are:

Natural (or casing-head)	Intermediate distillates	Waxes (candles, sealing, paper
Gasoline and natural gas	Heavy fuel oils	treating, insulating)
LPG	Diesel oils	Residues
Light distillates	Gas oils	Lubricating oil

Motor gasolines	Heavy distillates	Fuel oils
Solvent naphthas	Heavy mineral oils	Petrolatum
Jet fuel	(medicinal)	Road oils
Kerosene	Heavy flotation oils	Asphalts
Light heating oils	Lubricating oils	Coke

2. Products of Refining

Precursors of Petrochemicals. As markets change, there is constant alteration in the materials used for the manufacture of petrochemicals. Almost any synthesis desired can be brought about; the problem is to do it at low cost with the equipment available. In earlier times, acetylene was used extensively for making petrochemicals, but it is difficult to make and store, so ethylene has now become the principal raw material for further synthesis. Precursors are reactive materials usually made by breaking down larger molecules, called feedstocks. Ethylene is currently being made from LPG, naphtha, gas oil, diesel fuel, ethane, propane, and butane, with coal a possibility soon to be explored, and some testing of liquefied coal already completed. The principal precursors are:

Acetylene	Propylene	Benzene	Xylenes
Ethylene	Butene	Toluene	Naphthalene

Ethylene, manufacturing from distillates, natural gas, or gas liquids, is the largest volume organic material. The conditions for its manufacture lie somewhere between those usually thought of as refining and those encountered in chemical production. Extremely large plants are built and being built. Some plants have a production capacity as large as 7×10^8 kg/year.

Propylene is rarely produced except as a coproduct with ethylene. Steam cracking of ethylene produces most of it, and virtually all of it is used for polymer production. The remainder, used mostly for chemical production, comes from oil refinery fluid catalytic crackers. Refinery propylene is used mainly for alkylation.

Aromatics are usually thought of as coal-derived, but the amount from that source in 1980 was almost vanishingly small, 4 percent of the benzene, 0.9 percent of the toluene, and only 0.1 percent of the xylenes. Benzene can be made by dehydrogenation of cyclohexane or substituted cyclohexanes, by aromatization of methycyclopentane, and by demethylation of toluene or xylenes. The demand for aromatics is large and attention is being given to find catalysts to produce more BTX (benzene-toluene-xylene) for chemical and high-grade fuel use.

Naphthalene is used in smaller quantities than the lighter aromatics, but its consumption is far from trivial. Dealkylation of a selected reformate stream using chromate-aluminum carbide catalyst gives a product which is purified to be purer than that formed form coal tar.

Light Distillates. Aviation gasoline, (automobile) motor gasoline, naphthas, petroleum solvents, jet-fuel, and kerosene are the fractions generally regarded as light distillates. Any given refinery rarely makes all of them. Gasoline is the most important product, and around 45 percent of the crude processed now ends up as gasoline.

Intermediate Distillates. These include gas oil, light and heavy domestic furnace oils, diesel fuels, and distillates used for cracking to produce more gasoline. These distillates are used mainly for transportation fuels in heavy truck, railroad, small commercial boats, standby and peak-shaving power plants, farm equipment, and wherever diesels are used to produce power. Home heating furnaces use these distillates.

Heavy Distillates. These are converted into lubricating oils, heavy oils for a variety of fuel uses, waxes, and cracking stock.

Residues. Some constituents are simply not volatile enough to be distilled, even under vacuum. These include asphalt, residual fuel oil, coke, and petroleum. These difficult salable materials are by-products of the refining process, and while many are extremely useful, most are difficult to dispose of and are relatively unprofitable.

Petroleum-derived chemicals, commonly known as petrochemicals, are made from petroleum and natural gas. Production of some of these products is very large, and over 1000 organic chemicals are derived from petroleum. Examples are carbon black, butadiene, styrene, ethylene glycol, polyethylene, etc.

3. Processing or Refining

Refining involves two major branches, separation processes and conversion processes. Particularly in the field of conversion, there are literally hundreds of processes in use, many of them patented. Even in a given refinery running a single crude, daily changes to accommodate changing markets and changing parameters of the conversion apparatus take place. No refinery on any day will operate exactly as shown, but all refineries will operate along the basic lines indicated.

Separation Processes. The unit operations used in petroleum refining are the simple, usual ones, but the interconnections and interactions may be complex. Most major units are commonly referred to as stills. A crude still consists of heat exchangers, a furnace, a fractionating tower, steam strippers, condensers, coolers, and auxiliaries. There are usually working tanks for temporary storage at the unit; frequently there are treating tanks, used for improving the color and removing objectionable components, particularly sulfur; blending and mixing tanks; receiving and storage tanks for crude feed; a vapor recovery system; spill and fire control systems; and other auxiliaries. For the refinery as a whole, a boiler house and usually an electrical generating system are added. A control room with instruments to measure, record, and control, thus keeping track of material which permits heat and material balances, forms the heart of the system. One of the major functions of the instruments is to permit accurate accounting of the materials and utilities used.

Conversion Processes. About 70 percent of the crude processed is subjected to conversion processing both carbonium ion and free radical mechanisms occur. The presence of catalysts, the temperature, and pressure determine which type predominates. The following are examples of the more important basic reactions which occur: cracking or pyrolysis, polymerization, alkylation, hydrogenation, hydrocracking, isomerization, and reforming or aromatization.

Selected from "Shreve's Chemical Process Industries, 6th edition, N. Shreve, McGraw-Hill, 1993"

New Words

1. millennium [mi'leniəm] *n.* 一千年
2. tubular ['tjuːbjulə] *a.* 管的，管式的，由管构成的
3. still [stil] *n.* 蒸馏釜，蒸馏
4. salable ['seiləbl] *a.* 畅销的，销路好的
5. hydrocracking *n.* 加氢裂化，氢化裂解
6. isomerization *n.* 异构化（作用）
7. constituent [kəns'titjuənt] *n.* 组成，组分，成分
8. asphalt ['æsfælt] *n.* (地)沥青，柏油
9. straight run gasoline 直馏汽油
10. antiknock ['ænti'nɔk] *a.* 抗爆的，防爆的，抗震的
11. alkylation [ælki'leiʃən] *n.* 烷基化，烷基取代
12. butene ['bjuːtiːn] *n.* 丁烯
13. benzenoid *a.* 苯（环）型的
14. ethylbenzene *n.* 乙（基）苯
15. tetraethyl lead 四乙基铅
16. finishing ['finiʃiŋ] *n.* 精加工，最终加工

17. desalt [ˈdiːˈsɔːlt] vt. 脱盐
18. paraffin base crude　石蜡基石油
19. asphalt base crude　沥青基石油
20. naphthene base crude　环烷基石油
21. mixed base crude　混合基石油
22. differentiate [difəˈrenʃieit] v. 区分，区别；求微分，求导数
23. residue [ˈreizidjuː] n. 残余物，残渣，剩余物
24. casing-head gas　油井气，油田气
25. LPG=liquefied petroleum gas　液化石油气
26. light distillate　轻馏分
27. motor gasoline　动力汽油，车用汽油
28. solvent naphtha　溶剂石脑油
29. jet fuel　喷气式发动机燃料
30. light heating oil　轻质燃料油
31. intermediate distillate　中间馏分
32. diesel [ˈdiːzl] n. 内燃机，柴油机
33. diesel oil　柴油
34. gas oil　粗柴油，瓦斯油，汽油
35. heavy distillate　重馏分
36. heavy mineral oil　重质矿物油
37. medicinal [meˈdisinl] a. 药的，药用的；n. 药物，药品
38. heavy flotation oil　重质浮选油
39. lubricating oil　润滑油
40. petrolatum [petrəˈleitəm]　石蜡油，软石蜡，矿脂
41. road oil　铺路沥青
42. precursor [priˈkɜːsə]　产物母体，前身，先驱

43. butane [ˈbjuːtein] n. 丁烷
44. dehydrogenation [diːˌhaidrədʒˈneiʃən] n. 脱氢（作用）
45. aromatization n. 芳构化
46. demethylation n. 脱甲烷（作用）
47. methylcyclopentane n. 甲基环戊烷
48. trivial [ˈtriviəl] a. 普通的，不重要的，无价值的
49. chromate [ˈkrəumeit] n. 铬酸盐
50. aviation [eiviˈeiʃən] n. 航空，飞行
51. standby [ˈstændbai] a. 备用的，后备的；n. 备用设备
52. carbon black　炭黑
53. glycol [ˈglaikɔl] n. 乙二醇
54. patent [ˈpeitənt] n. 专利，专利权；vt. 取得……的专利权
55. parameter [pəˈræmitə] n. 参数，系数
56. fractionate [ˈfrækʃəneit] vt. 使分馏，把……分成几部分
57. stripper [ˈstripə] n. 汽提塔，解吸塔
58. carbonium n. 阳碳，正碳
59. carbonium ion mechanism　正碳离子机理
60. radical [ˈrædikəl] n. 基，原子团；根部；根式
61. free radical mechanism　自由基机理
62. predominate [priˈdɔmineit] vi. 占优势，居支配地位
63. pyrolysis [paiˈrɔlisis] n. 热解（作用），高温分解
64. hydrogenation [haiˌdrɔdʒiˈneiʃən] n. 加氢（作用）

Lesson 19
Chemical Industry and Environment

How can we reduce the amount of waste that is produced? And how can we close the loop by redirecting spent materials and products into programs of recycling? All of these questions must be answered through careful research in the coming years as we strive to keep civilization in balance with nature.

1. Atmospheric Chemistry

Coal-burning power plants, as well as some natural processes, deliver sulfur compounds to the stratosphere[1], where oxidation produces sulfuric acid particles that reflect away some of the incoming visible solar radiation. In the troposphere[2], nitrogen oxides produced by the combustion of fossil fuels combine with many organic molecules under the influence of sunlight to produce urban smog. The volatile hydrocarbon isoprene, well known as a building block of synthetic rubber, is also produced naturally in forests. And the chlorofluorocarbons, better known as CFCs, are inert in automobile air conditioners and home refrigerators but come apart under ultraviolet bombardment in the mid-stratosphere with devastating effect on the earth's stratospheric ozone layer. The globally averaged atmospheric concentration of stratospheric ozone itself is only 3 parts in 10 million, but it has played a crucial protective role in the development of all biological life through its absorption of potentially harmful short-wavelength solar ultraviolet radiation.

During the past 20 years, public attention has been focused on ways that mankind has caused changes in the atmosphere: acid rain, stratospheric ozone depletion, greenhouse warming, and the increased oxidizing capacity of the atmosphere. We have known for generations that human activity has affected the nearby surroundings, but only gradually have we noticed such effects as acid rain on a regional then on an intercontinental scale. With the problem of ozone depletion and concerns about global warming, we have now truly entered an era of global change, but the underlying scientific facts have not yet been fully established.

2. Life Cycle Analysis

Every stage of a product's life cycle has an environmental impact, starting with extraction of raw materials, continuing through processing, manufacturing, and transportation, and concluding with consumption and disposal or recovery. Technology and chemical science are challenged at every stage. Redesigning products and processes to minimize environmental impact requires a new philosophy of production and a different level of understanding of chemical transformations. Environmentally friendly products require novel materials that are reusable, recyclable, or biodegradable[3]; properties of the materials are determined by the chemical composition and structure. To minimize waste and polluting by-products, new kinds of chemical process schemes will have to be developed. Improved chemical separation techniques are needed to enhance efficiency and to remove residual pollutants, which in turn will require new chemical treatment methods in order to render them harmless. Pollutants such as radioactive elements and toxic heavy metals that cannot be readily converted into harmless materials will need to be immobilized in inert materials so that they can be safely stored. Finally, the leftover pollution of an earlier, less environmentally aware era demands improved chemical and biological remediation techniques.

Knowledge of chemical transformations can also help in the discovery of previously unknown environmental problems. The threat to the ozone layer posed by CFCs was correctly anticipated through fundamental studies of atmospheric chemistry, eventually leading to international agreements for phasing out the production of these otherwise useful chemicals in favor of equally functional but environmentally more compatible alternatives. On the other hand, the appearance of the ozone hole over the Antarctic came as a surprise to scientists and only subsequently was traced to previously unknown chlorine reactions occurring at the surface of nitric acid crystals in the frigid Antarctic stratosphere. Thus it is critically important to improve our understanding of the chemical processes in nature, whether they occur in fresh water, saltwater, soil, subterranean environments, or the atmosphere.

3. Manufacturing with Minimal Environmental Impact

Discharge of waste chemicals to the air, water, or ground not only has a direct environmental impact, but also constitutes a potential waste of natural resources. Early efforts to lessen the environmental impact of chemical processes tended to focus on the removal of harmful materials form a plant's waste stream before it was discharged into the environment. But this approach addresses only half of the problem; for an ideal chemical process, no harmful by-products would be formed in the first place. Any discharges would be at least as clean as the air and water that were originally taken into the plant, and such a process would be "environmentally benign."

Increasing concern over adverse health effects has put a high priority on eliminating or reducing the amounts of potentially hazardous chemicals used in industrial processes. The best course of action is to find replacement chemicals that work as well but are less hazardous. If a substitute cannot be found for a hazardous chemical, then a promising alternative strategy is to develop a process for generating it on-site and only in the amount needed at the time.

Innovative new chemistry has begun delivering environmentally sound processes that use energy and raw materials more efficiently. Recent advances in catalysis, for example, permit chemical reactions to be run at lower temperatures and pressures. This change, in turn, reduces the energy demands of the processes and simplifies the selection of construction materials for the processing facility. Novel catalysts are also being used to avoid the production of unwanted by-products.

4. Control of Power Plant Emissions

Coal-, oil-, and natural-gas-fired power generation facilities contribute to the emissions of carbon monoxide, hydrocarbons, nitrogen oxides, and a variety of other undesired by-products such as dust and traces of mercury. A rapidly increasing array of technologies is now available to reduce the emissions of unwanted species to meet national or local standards. Chemists and chemical engineers have made major contributions to the state of the art, and catalytic science is playing a critical role in defining the leading edge.

The simultaneous control of more than one pollutant is the aim of some recently developed catalyst or sorbent technologies. For example, catalytic methods allow carbon monoxide to be oxidized at the same time that nitrogen oxides are being chemically reduced in gas turbine exhaust. Other research efforts are aimed at pilot-plant evaluation of the simultaneous removal of sulfur and nitrogen oxides form flue gas by the action of a single sorbent and without the generation of massive volumes of waste products.

5. Environmentally Friendly Products

Increased understanding of the fate of products in the environment has led scientists to design "greener" products. A significant early example comes from the detergent industry in the 1940s and 1950s, new products were introduced that were based on synthetic surfactants called branched alkylbenzene sulfonates. These detergents had higher cleaning efficiency, but it was subsequently discovered that their presence in waste water caused foaming in streams and rivers. The problem was traced to the branched alkylbenzene sulfonates; unlike the soaps used previously, these were not sufficiently biodegraded by the microbes in conventional sewage treatment plants. An extensive research effort to understand the appropriate biochemical processes permitted chemists to design and synthesize another new class of surfactants, linear alkylbenzene sulfonates. The similarity in molecular structure between these new compounds and the natural fatty acids of traditional soaps allowed the microorganisms to degrade the new formulations, and the similarity to the branched alkylbenzene sulfonates afforded outstanding detergent performance.

Novel biochemistry is also helping farmers reduce the use of insecticides. Cotton plants, for example, are being genetically modified to make them resistant to the cotton bollworm. A single gene from a naturally occurring bacterium, when transferred into cotton plants, prompts the plant to produce a protein that is ordinarily produced by the bacterium. When the bollworm begins to eat the plant, the protein kills the insect by interrupting its digestive processes.

6. Recycling

Increasing problems associated with waste disposal have combined with the recognition that some raw materials exist in limited supply to dramatically increase interest in recycling. Recycling of metals and most paper is technically straightforward, and these materials are now commonly recycled in many areas around the world. Recycling of plastics presents greater technical challenges. Even after they are separated from other types of waste, different plastic materials must be separated form each other. Even then, the different chemical properties of the various types of plastic will require the development of a variety of recycling processes.

Some plastics can be recycled by simply melting and molding them or by dissolving them in an appropriate solvent and then reformulating them into a new plastic material. Other materials require more complex treatment, such as breaking down large polymer molecules into smaller subunits that can subsequently be used as building blocks for new polymers. Indeed, a major program to recycle plastic soft drink bottles by this route is now in use.

A great deal of research by chemists and chemical engineers will be needed to successfully develop the needed recycling technologies. In some cases, it will be necessary to develop entirely new polymers with molecular structures that are more amenable to the recycling process.

7. Separation and Conversion for Waste Reduction

New processes are needed to separate waste components requiring special disposal from those that can be recycled or disposed of by normal means. Development of these processes will require extensive research to obtain a fundamental understanding of the chemical phenomena involved.

Metal-Bearing Spent Acid Waste. Several industrial processes produce acidic waste solutions in large quantities. Could this waste be separated into clean water, reusable acid, and a sludge from which the metals could be recovered? Such processes would preserve the environment, and their costs could be competitive with disposal costs and penalties.

Industrial Waste Treatment. The hazardous organic components in industrial wastewater could be destroyed with thermocatalytic or photocatalytic processes. A promising line of research employs "supercritical" water at high temperatures and pressures. Under these conditions, water exhibits very different chemical and physical properties. It dissolves and allows reactions of many materials that are nearly inert under normal conditions.

High-Level Nuclear Waste. Substantial savings would be achieved if the volume and complexity of nuclear waste requiring storage could be significantly reduced; this reduction would require economic separation of the radioactive components from the large volumes of other materials that accompany the nuclear waste. The hazardous chemical waste might then be disposed of separately. The disposal of nuclear waste will require major research and development efforts over many years.

Membrane Technology. Separations involving semipermeable membranes offer considerable promise. These membranes, usually sheets of polymers, are impervious to some kinds of chemicals but not to others. Such membranes are used to purify water, leaving behind dissolved salts and providing clean drinking water. Membrane separation techniques also permit purification of wastewater from manufacturing. Membrane separations are also applicable to gases and are being used for the recovery of minor components in natural gas, to enhance the heating value of natural gas by removal of carbon dioxide, and for the recovery of nitrogen from air. Research challenges include the development of membranes that are chemically and physically more resilient, that are less expensive to manufacture, and that provide better separation efficiencies to reduce processing costs.

Biotechnology. Scientists have turned to nature for help in destroying toxic substances. Some microorganisms in soil, water, and sediments can adapt their diets to a wide variety of organic chemicals; they have been used for decades in conventional waste treatment systems. Researchers are now attempting to coax even higher levels of performance from these gifted microbes by carefully determining the optimal physical, chemical, and nutritional conditions for their existence. Their efforts may lead to the design and operation of a new generation of biological waste treatment facilities. A major advance in recent years is the immobilization of such microorganisms in bioreactors, anchoring them in a reactor while they degrade waste materials. Immobilization permits high flow rates that would flush out conventional reactors, and the use of new, highly porous support materials allows a significant increase in the number of microorganisms for each reactor.

Selected from "Critical Technologies: the Role of Chemistry and Chemical Engineering National Research Council, National Academy of Sciences, 1992"

New Words

1. strive [straiv] *vi.* 努力，奋斗，力求；斗争，反抗
2. stratosphere ['strætəusfiə] *n.* 同温层，平流层
3. troposphere ['trɔpəsfiə] *n.* 对流层
4. isoprene ['aisəupri:n] *n.* 异戊二烯
5. building block 结构单元，预制件，积木
6. chlorofluorocarbon *n.* 含氯氟烃(CFC)
7. ultraviolet [,ʌltrə'vaiələt] *a.* 紫外的，紫外线的
8. bombardment [bɔm'bɑ:dmənt] *n.* 照射，辐照；轰击，打击

Lesson 19 Chemical Industry and Environment

9. devastate ['devəsteit] *vt.* 破坏，毁坏，使荒芜
10. ozone ['əuzəun] *n.* 臭氧
11. greenhouse ['gri:nhaus] *n.* 温室，暖房
12. intercontinental [,intəkɔnti'nentl] *a.* 洲际的
13. biodegradable *a.* 可生物降解的
14. leftover *a.; n.* 剩余的（物）
15. remediation *n.* 补救，修补；治疗
16. pose [pəuz] *v.* 造成，形成，提出
17. compatible [kəm'pætəbl] *a.* 兼容的，可共存的；一致的，相似的，协调的
18. Antarctic [ænt'ɑ:ktik] *n.; a.* 南极地带（的）
19. frigid ['fridʒid] *a.* 寒冷的，严寒的
20. subterranean [,sʌbtə'reiniən] *a.* 地下的，隐藏的，秘密的
21. benign [bi'nain] *a.* 有益于健康的，良好的，[医] 良性的
22. on-site *a.* （在）现场的，就地的
23. sorbent ['sɔ:bənt] *n.* 吸附剂，吸收剂
24. alkylbenzene sulfonate 烷基苯磺酸盐
25. microbe ['maikrəub] *n.* 微生物，细菌
26. insecticide [in'sektisaid] *n.* 杀虫剂，农药
27. bollworm *n.* 蝼蛉
28. insect ['insekt] *n.* 昆虫
29. digestive [di'dʒestiv] *a.* 消化的，助消化的
30. subunit ['sʌb,ju:nit] *n.* 副族，子单元，亚组，子群
31. sludge [slʌdʒ] *n.* 淤泥，泥状沉积物；淤渣
32. penalty ['penlti] *n.* 罚款；损失
33. impervious [im'pɜ:vjəs] *a.* 不能透过的，不可渗透的
34. resilient [ri'ziliənt] *a.* 有弹性的，能恢复原状的
35. sediment ['sedimənt] *n.* 沉积物；沉积，沉淀
36. diet ['daiət] *n.* 食物，饮食
37. coax [kəuks] *vt.* 耐心地处理，慢慢地把……弄好；诱，哄
38. immobilization [i'məubilai'zeiʃən] *n.* 固定，定位，降低流动性

Notes

1. stratosphere:平流层。对流层顶(tropopause)以上到离地约50km的大气层，因大气多平流运动，故名。其层内温度一般随高度增加而升高。
2. troposphere:对流层。大气的低层，对流运动显著的气层。其厚度由地球表面向上至不同的高度，极地约9km，赤道约17km。层内的温度随高度颇有规则地降低。
3. biodegradation:生物降解。物质被细菌分解。

Exercises

1. Put the following into Chinese:
   stratosphere   troposphere   CFC   bombardment
   devastate   remediation   on-site   microbe
   insecticide   coax   hazard   ingredient

2. Put the following into English:
   紫外的   臭氧   可生物降解的   烷基苯磺酸盐
   污水   温室效应   吸附剂   剩余的
   研究所   兼容的   淤泥   保障

3. Comprehension and toward interpretation
   a. What is the character of chemical industry?
   b. What is the relation between the process safety and the growth of chemical industry?
   c. What is cluded in the technology of safety?

d. Chemical process safety is the most important one, why ?

Reading Material

Chemical Process Safety

In 1987, Robert M. Solow, an economist at the Massachusetts Institute of Technology, received the Nobel Prize[1] in economics for his work in determining the sources of economic growth. Professor Solow concluded that the bulk of an economy's growth is the result of technological advances.

It is reasonable to conclude that the growth of an industry is also dependent on technological advances. This is especially true in the chemical industry, which is entering an era of more complex processes: higher pressure, more reactive chemicals, and exotic chemistry[2].

More complex processes require more complex safety technology. Many industrialists even believe that the development and application of safety technology is actually a constraint on the growth of the chemical industry. As chemical process technology becomes more complex, chemical engineers will need a more detailed and fundamental understanding of safety. H. H. Fawcett has said that to know is to survive and to ignore fundamentals is to court disaster.

Since 1950, significant technological advances have been made in chemical process safety. Today, safety is equal in importance to production and has developed into a scientific discipline which includes many highly technical and complex theories and practices. Examples of the technology of safety include:

a) Hydrodynamic models representing two-phase flow through a vessel relief[3].

b) Dispersion models representing the spread of toxic vapor through a plant after a release[4].

c) Mathematical techniques to determine the various ways that processes can fail, and the probability of failure[5].

Recent advances in chemical plant safety emphasize the use of appropriate technological tools to provide information for making safety decisions with respect to plant design and operation. The word *safety* used to mean the older strategy of accident prevention through the use of hard hats[6], safety shoes, and a variety of rules and regulations. The main emphasis was on worker safety. Much more recently, *safety* has been replaced by *loss prevention*[7]. This term includes hazard identification, technical evaluation, and the design of new engineering features to prevent loss. The words *safety* and *loss prevention* will be used synonymously throughout for convenience.

Safety, hazard, and risk are frequently-used terms in chemical process safety. Their definitions are:

a) *Safety* or *loss prevention* is the prevention of accidents by the use of appropriate technologies to identify the hazards of a chemical plant and to eliminate them before an accident occurs.

b) A *hazard* is anything with the potential for producing an accident.

c) *Risk* is the probability of a hazard resulting in an accident.

Chemical plants contain a large variety of hazards. First, there are the usual mechanical harzards that cause worker injuries from tripping, failing, or moving equipment. Second, there are chemical hazards, these include fire and explosion hazards, reactively hazards, and toxic hazards.

As will be shown later, chemical plants are the safest of all manufacturing facilities. However, the potential always exists for an accident of catastrophic proportions. Despite substantial safety programs by the chemical industry, headlines of the type shown in Figure continue to appear in the newspapers.

A successful safety program requires several ingredients. These ingredients are:
a) Safety knowledge
b) Safety experience
c) Technical competence
d) Safety management support
e) Commitment.

Selected from "English for Chemical Engineers, by Ma Zhengfei etc., Southeast University Press, 2006, 85-86".

New Words

1. bulk [bʌlk]   *n.* 块，主体
2. hazard ['hæzəd]   *n.* 危害
3. constraint [kən'streint]   *n.* 强制，制约，限制
4. synonymously [si'nɔniməsli]   *ad.* 同（意）义地
5. eliminate [i'limineit]   *n.* 除去，排除，消除
6. facility [fə'siliti]   *n.* 工厂，研究所，实验室，装备
7. catastrophic [,kætə'strɔfik]   *a.* 灾难性的，灾变的
8. ingredient [in'griːdiənt]   *n.* 组成部分，成分
9. competence ['kɔmpitəns]   *n.* 能力
10. support [sə'pɔːt]   *n.* 保障
11. commitment [kə'mitmənt]   *n.* 承担义务

Notes

1. Nobel Prize 诺贝尔奖。
2. exotic chemistry 奇异化学，即奇异原子化学（exotic atom chemistry），包括正子素化学、介子素化学、介子原子化学等。
3. Hydrodynamic models representing two-phase flow through a vessel relief. 参考译文：表示通过减压容器的两相流水力学模型。
4. Dispersion models representing the spread of toxic vapor through a plant after a release. 参考译文：表示经工厂排放的有毒蒸气散播分散模型。
5. Mathematical techniques to determine the various ways that processes can fail, and the probability of failure. 参考译文：通过过程可能失败的各种途径和失败概率的数学技术。
6. hard hat 保护帽。
7. loss prevention 损失预防。

Lesson 20
Vapor-Phase Chromatography

Vapor-phase chromatography (V>P>C), also known as gas chromatography or gas-liquid partition chromatography (G>L>P>C), is used extensively in the separation and identification of mixtures of volatile compounds. It is also one of the most useful techniques for the quantitative determination of the components of a mixture.

In gas chromatography each component of a volatile mixture of compounds is partitioned between vapor and liquid phases as the mixture is passed through a column containing a suitable, relatively non-volatile liquid (the adsorbent) impregnated in a finely divided, inert, porous solid material (the solid support) .The various components of the mixture pass through the column at different rates, and each component can be detected and, if desired, collected as it emerges from the column.

In actual practice, a very small sample of the mixture is injected into a heated chamber where it is vaporized and swept through the column (which is encased in an oven which also is usually heated) with the aid of an inert carrier gas, most commonly helium. After the components of the mixture have been separated as they pass through the column, they pass through a detector located at the end of the column and are recorded as peaks on a mechanical recorder. A schematic diagram of a gas chromatograph is shown in Figure 20-1.

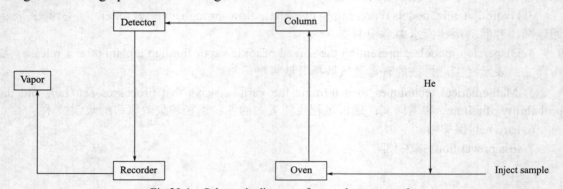

Fig.20-1 Schematic diagram of a gas chromatograph

The time required for a compound to pass through a column is known as its retention time, which depends on many variables, including the following.
1. The nature of the adsorbent.
2. The concentration of adsorbent in the inert solid support.
3. The degree to which the compound being analyzed is adsorbed by the adsorbent.
4. The volatility of the compound.
5. The column temperature.
6. The rate of flow of carrier gas.

7. The nature of the carrier gas.

8. The dimensions of the column.

By suitable adjustments of these variables it is usually possible to effect a clean separation of all of the components of a mixture of compounds. Some of the liquid adsorbents in common use are silicone rubbers, apiezon greases, waxes, oils, carbowaxes and dialkyl phthalates. Some of the inert, porous solid supports for the adsorbents are firebrick, Celite and Teflon. Nitrogen or argon is sometimes used as the carrier gas in place of helium.

The detector must be capable of measuring some difference between the carrier gas alone and the carrier gas mixed with an effluent compound. Differences in thermal conductivity are easy to measure and provide a high sensitivity to detection. Thus, this type of detector is the one most frequently used. Changes in the thermal conductivity of the effluent gases are sensed by the detector and relayed to a recorder by electronic means. The recorder provides a display of peaks as a function of time. Each peak represents a new compound that has passed through the detector. Furthermore, the area under each peak, when suitably calibrated, represents a measure of the amount of the compound present in the original mixture.

As mentioned above, the retention time of a compound under a specified set of conditions is a physical constant that can be used in qualitative analysis. For example, you might wish to use p-xylene as a reagent, but might suspect that a commercial sample of the compound is contaminated with small amounts of o-xylent and m-xylene. If your suspicion is correct, you will find that a vapor-phase. You can then confirm your suspicion as to the nature of one of the contaminants by deliberately adding some o-xylene to the reagent and obtaining a new vapor-phase chromatogram. If the contaminant in the reagent is indeed o-xylene, one of the minor peaks will be larger in the second chromatogram than in the first. The same type of control experiment can then be carried out with addition of m-xylene to the reagent. Alternatively, the retention times of pure o-xylene and m-xylene can be obtained under the same condition as used for the reagent. The retention times of the two impurities present in the p-xylene should correspond to those of pure o-xylene and m-xylene, respectively, if these are the impurities. Reasonable care must be exercised in the interpretation of vapor-phase chromatograms for purposes of qualitative analysis. Just as two different solid compounds may possess the same or very nearly the same melting point, two different volatile compounds may exhibit identical retention times under a given set of conditions. Thus, it is usually wise to confirm a tentative identification of a compound through its retention time by some other means. One of the best ways to do this is to condense the compound as it leaves the detector unit of the gas chromatograph, take the infrared spectrum of the condensate and then compare it with the spectrum of the compound whose presence is suspected.

The best use of vapor-phase chromatography is in the quantitative analysis of mixtures of volatile compounds. As a first approximation, it can be assumed that the peak area of a given compound in its vapor-phase chromatogram will be directly proportional to its weight percent in the original mixture. However, this is not strictly true, and suitable calibration curves must be constructed to derive accurate values. A variety of methods can be used to measure the peak areas of the components of a mixture. One sophisticated method is to use an electronic integrator connected to the recorder. However, while this approach is routine in nuclear magnetic resonance spectral determinations there are relatively few commercial vapor-phase fractometers that are equipped with an integrator. Since the recorder paper used in vapor-phase chromatography has a very uniform thickness and density, a simple way to calculate relative area is to cut out the peaks and weigh the pieces of paper on an analytical grade balance. Alternatively, you may use a planimeter to measure the area under each peak. A fourth method is to calculate the area under each peak by use of the relationship:

$$Area = Peak\ Height \times Width\ at\ half\ Height$$

It should be kept in mind that this method can be used satisfactorily only when the peaks are symmetrical and well separated from one another.

Selected from "Specialized Chemical English (unformal published) (volumn two), by Cui Bo etc., 1994, 133-139"

New Words

1. vapor-phase chromatography ['veipə'feizkroumə'tɔgrəfi] 气相色谱（法）
2. chromatography [kroumə'tɔgrəfi] n. 层析法，色谱法
3. gas chromatography 气体色谱（法）
4. partition [pɑː'tiʃən] n.; vt. 分离，分开，分配
5. gas-liquid par tition chromatography 气液分配色谱法
6. separate ['sepəreit] v. 分离，分开
7. determination [ditəːmi'neiʃən] n. 测定
8. identification [aidentifi'keiʃən] n. 鉴别，鉴定
9. adsorbent [æd'sɔːbənt] n. 吸附剂
10. impregnate ['impregneit] vt. 浸渍
11. porous ['pɔːrəs] a. 多孔的
12. collect [kə'lekt] vt. 收集
13. emerge [iməːdʒ] vi. 出现，形成
14. practice ['præktis] n. 实践操作
15. sample ['sɑːmpl] n. 试样
16. inject [in'dʒekt] vt. 注射
17. chamber ['tʃeimbə] v. 室，房间
18. vaporize ['veipəraiz] v. 汽化
19. encase [in'keis] vt. 把……放入套内
20. oven ['ʌvn] n. 炉，烘箱
21. carrier gas ['kæriəgæs] n. 载气
22. detector [di'tektə] n. 检测器
23. recorder [ri'kɔːdə] n. 记录器
24. schematic diagram [ski'mætik'daiəgræm] 示意图
25. retention time [ri'tenʃəntaim] 保留时间
26. degree [di'griː] n. 程度，度
27. analyze ['ænəlaiz] vt. 分析
28. adsorb [əd'sɔːb] vt. 吸附
29. volatility [vɔlə'tiliti] n. 挥发度，挥发性
30. dimension [di'menʃən] vt. 尺寸
31. effect [i'fekt] vt. 实现; n. 效应，影响，效果
32. clean [kliːn] a. 彻底的，完全的
33. silicone rubber [silikən'rʌbə] 硅橡胶
34. apiezon [ə'piːzɔn] n. 阿匹松，饱和烃
35. phthalate [f'θæleit] n. 邻苯二甲酸盐（酯）
36. dialkyl [dai'ælkil] phthalate n. 邻苯二甲酸二烷基酯
37. firebrick ['faiəbrik] n. 耐火砖
38. celite [si'lait] n. 硅藻土
39. Teflon ['teflɔn] n. 特氟隆，聚四氟乙烯
40. effluent ['efluənt] n. 流出物；a. 流出的
41. sensitivity [sensi'tiviti] n. 敏感度，灵敏度
42. sense [sens] vt. (自动)检测
43. relay ['riːlei] vt 传递
44. display [dis'plei] vt.; n. 显示，表现
45. calibrate ['kælibreit] vt. 校准
46. detector signal [signl] 检测器信号
47. methyl ester 甲酯
48. fatty acid ['fæti'æsid] 脂肪酸
49. constant ['kɔnstənt] n. 常数
50. suspect [səs'pekt] vt. 猜想，怀疑
51. suspicion [səs'piʃən] n. 猜想，怀疑
52. chromatogram ['kroumətəgræm] n. 色谱图
53. major ['meidʒə] a. 主要的
54. minor ['mainə] a. 较小的，较次要的
55. contaminant [kən'tæminənt] n. 沾污物，污染物
56. deliberately [di'libərətli] ad. 故意地，审

慎地
57. control [kən'trəul] n. 对照（物），控制
58. wise [waiz] a. 明智的，考虑周到的，慎重的

59. confirm [kən'fə:m] vt. 进一步证实，确定
60. tentative ['tentətiv] a. 不明确的
61. condense [kən'dens] v. 冷凝，缩合
62. infrared ['infrə'red] a. 红外线的

Exercises

1. Put the following into Chinese:

chromatography	collect	inject	minor
partition	emerge	firebrick	infrared
impregnate	dimension	sensitivity	stationary
loop	eluent	labile	valve

2. Put the following into English:

气相色谱	多孔的	常熟	试样
鉴别	挥发性	冷凝	固定的
吸附剂	吸附	载气	衍生法
缓冲液	稳定器	密封	筛分

Reading Material

High Performance Liquid Chromatography and Capillary Electrophoresis

High Performance Liquid Chromatography.

Liquid Chromatography was first discovered in 1903 by M. S. Tswett, who used a chalk column to separate the pigments of green leaves[1]. Only in 1960's the more and more emphasis was placed on the development of liquid chromatography. High Performance Liquid Chromatography (HPLC) is one mode of chromatography, the most widely used analytical technique. Chromatographic processes can be defined as separation techniques involving mass-transfer between stationary and mobile phases. HPLC utilizes a liquid mobile phase to separate the components of a mixture. These components (or analytes) are first dissolved in a solvent, and then forced to flow through a chromatographic column under a high pressure. In the column, the mixture is resolved into its components. The amount of resolution is important, and is dependent upon the extent of interaction between the solute components and the stationary phase. The stationary phase is defined as the immobile packing material in the column. The interaction of the solute with mobile and stationary phases can be manipulated through different choices of both solvents and stationary phases. As a result, HPLC acquires a high degree of versatility not found in other chromatographic systems and it has the ability to easily separate a wide variety of chemical mixtures.

The basic components of an HPLC system include a solvent reservoir, pump, injector, analytical column, detector, recorder and waste reservoir. Other important elements are an inlet solvent filter, post-pump inline filter, sample filter, precolumn filter, guard column, back-pressure regulator and/or solvent sparging system. The function of each of these components is briefly described in Fig.20-2.

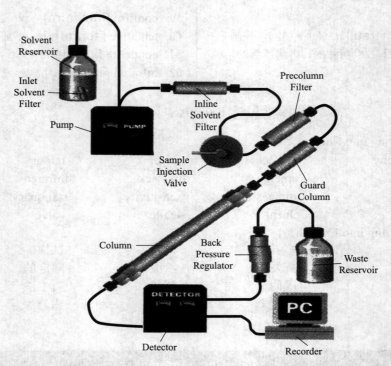

Fig. 20-2 The function of each of these components

An HPLC system begins with the solvent reservoir, which contains the solvent used to carry the sample through the system. The solvent should be filtered with an inlet solvent filter to remove any particles that could potentially damage the system's sensitive components. Solvent is propelled through the system by the pump. This often includes internal pump seals, which slowly break down over time. As these seals break down and release particles into the flow path, an inline solvent filter prevents any post-pump component damage. The next component in the system is the sample injector, also known as the injection valve. This valve, equipped with a sample loop of the appropriate size for the analysis being performed, allows for the reproducible introduction of sample into the flow path. Because the sample often contains particulate matter, it is important to utilize either a sample filter or a precolumn filter to prevent valve and column damage. Following the injector, an analytical column allows the primary sample separation to occur. This is based on the differential attraction of the sample components for the solvent and the packing material within the column. However, a sacrificial guard column is often included just prior to the analytical column to chemically remove components of the sample that would otherwise foul the main column[2]. Following the analytical column, the separated components pass through a detector flow cell before they pass into the waste reservoir. The sample components' presence in the flow cell prompts an electrical response from the detector, which is digitized and sent to a recorder. The recorder helps analyze and interpret the date. As a final system enhancement, a back pressure regulator is often installed immediately after the detector. This device prevents solvent bubble formation until the solvent is completely through the detector. This is important because bubbles in a flow cell can interfere with the detection of sample components. Alternatively, an inert gas sparging system may be installed to force dissolved gasses out of the solvent being stored in the solvent reservoir.

Let us consider a separation of a two component mixture dissolved in the eluent. Assume that component A has the same interaction with the adsorbent surface as an eluent, and component B has strong excessive interaction[3]. Being injected into the column, these components will be forced through by eluent flow. Molecules of component A will interact with the adsorbent surface and retard on it by the same way as an eluent molecules. Thus, as an average result, component A will move through the column with the same speed as an eluent.

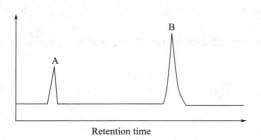

Fig. 20-3 The general shape of the chromatogram for this mixture

Molecules of the component B being adsorbed on the surface (due to their strong excessive interactions) will sit on it much longer. Thus, it will move through the column slower than the eluent flow.

Fig. 20-3 represents the general shape of the chromatogram for this mixture.

Usually a relatively narrow band is injected (5~20 μL injected volume). During the run, the original chromatographic band will be spread due to the nonevent flows around and inside the porous particles, slow adsorption kinetics, longitudinal diffusion, and other factors. These processes together produce so called band broadening of the chromatographic zone. In general, the longer the component retained on the column, the troader its zone (peak on the chromatogram). Separation performance depends on both component retention and band broadening. Band broadening is, in general, a kinetic parameter, dependent on the adsorbent particle size, porosity, pore size, column size, shape, and packing performance. On the other hand, retention does not depend on the above mentioned parameters, but it reflects molecular surface interactions and depends on the total adsorbent surface.

Today, HPLC is the most widely used analytical separation method. The method is popular because it is non-destructive and may be applied to thermally labile compounds (unlike GC); it is also a very sensitive technique since it incorporates a wide choice of detection methods[4]. With the use of post-column derivatization methods to improve selectivity and detection limits, HPLC can easily be extended to trace determination of compounds that do not usually provide adequate detector response[5]. The wide applicability of HPLC as a separation method makes it a valuable separation tool in many scientific fields.

Liquid chromatography is useful for a multitude of applications in industry and academia. Its use can be broken down into two classifications, analytical LC and preparative LC. In analytical LC the goal is identification and quantification of given components within a sample, usually in the picogram to milligram range. In preparative LC, the objective is to isolate or collect the separated components of the sample in the mg to kg range.

In industry, as well as the sciences analytical LC is employed for:
- Basic Research
- Quality Assurance
- Methods Development

Liquid chromatography is widely used in different types of industry:
- Life Science—proteins, nucleic acids, carbohydrates, lipids, metabolites
- Pharmaceuticals

- Biotechnology
- Industrial Chemicals—fine chemicals, polymers, synthetic mixtures
- Food and Agriculture Processing—plant products, agrochemicals

The preparative capabilities of LC are applied at different levels or scales of isolation:
- Small-scale or semi-preparative(mg to g)
- Pilot-scale (g to kg)
- Production or process scale (kg to ton)

In preparative LC, a partial list of specific needs or reasons to isolate or purify samples is as follows:
- Drug Efficiency Studies
- Full-scale Production of Drugs (e.g., interferon, insulin)
- Spectroscopy /Structure Elucidation
- Biological Screening
- Physical Testing

High Performance Capillary Electrophoresis.

Electrophoresis refers to the migration of charged electrical species when dissolved, or suspended, in an electrolyte through which an electric current is passed. Cations migrate toward the negatively charged electrode (cathode) and anions are attracted toward the positively charged electrode (anode). Neutral solutes are not attracted to either electrode. Conventionally electrophoresis has been performed on layers of gel or paper. The traditional electrophoresis equipment offered a low level of automation and long analysis times. Detection of the separated bands was performed by post-separation visualization. The analysis times were long as only relatively low voltages could be applied before excessive heat formation caused loss of separation.

The advantages of conducting electrophoresis in capillaries were highlighted in the early 1980's by the work of Jorgenson and Lukacs who popularized the use of CE. Performing electrophoretic separations in capillaries was shown to offer the possibility of automated analytical equipment, fast analysis times and on-line detection of the separated peaks. Heat generated inside the capillary was effectively dissipated through the walls of the capillary which allowed high voltages to be used to achieve rapid separations. The capillary was inserted through the optical center of a detector which allowed on-capillary detection.

Capillary electrophoresis has grown to become a collection of a range of separation techniques which involve the application of high voltages across buffer filled capillaries to achieve separations. The variations include separation based on size and charge differences between analytes (termed Capillary Zone Electrophoresis, CZE, or Free Solution CE, FSCE), separation of neutral compounds using surfactant micelles (Micellar electrokinetic capillary chromatography, MECC or sometimes referred to as MEKC), sieving of solutes through a gel network (Capillary Gel Eletrophoresis, CGE), and separation of zwitterionic solutes within a pH gradient (Capillary Isoelectric Focusing, CIEF)[6]. Capillary electrochromatography (CEC) is an associated electrokinetic separation technique which involves applying voltages across capillaries filled with silica gel stationary phases. Separation selectivity in CEC is a combination of both electrophoretic and chromatographic processes. Many of the CE separation techniques rely on the presence of an electrically induced flow of solution eletroosmotic flow, EOF within the capillary to pump solutes towards the detector. The basis to EOF is discussed later in this chapter.

FSCE and MECC are the most frequently used separation techniques in pharmaceutical analysis. GCE and CIEF are of importance for the separation of biomolecules such as DNA and proteins respectively and are becoming of increasing importance as development of biotechnology derived drugs is becoming more frequent. Generally, CE is performed using aqueous based electrolytes, however there is a growing use of non-aqueous solvents in CE.

Operation of a CE system involves application of a high voltage (typically 10~30kV) across a narrow bore (25~100 μm) capillary. The capillary is filled with electrolyte solution which conducts current through the inside of the capillary. The ends of the capillary are dipped into reservoirs filled with the electrolyte. Electrodes made of an inert material such as platinum are also inserted into the electrolyte reservoirs to complete the electrical circuit. A small volume of sample is injected into one end of the capillary. The capillary passes through a detector, usually a UV absorbance detector, at the opposite end of the capillary. Application of a voltage causes movement of sample ions towards their appropriate electrode usually passing through the detector. The plot of detector response with time is generated which is termed an electropherogram. A flow of electrolyte, known as electroosmotic flow, EOF, results in a flow of the solution along the capillary usually towards the detector. This flow can significantly reduce analysis times or force an ion to overcome its migration tendency towards the electrode it is being attracted to by the sign of its charge.

Detailed treatments of the background theory and non-pharmaceutical based applications can be obtained from a number of reference books.

Commercially available CE instruments (Fig.20-4) are PC controlled and consist of a buffer filled capillary passing through the optical center of a detector, a means of introducing the sample into the capillary, a high voltage power supply and an autosampler.

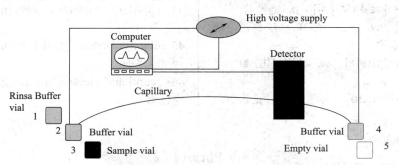

Fig.20-4 Typical CE separation system

The typical voltages used are in the range of 5~30kV which results in currents in the range of 10~100 μA. Higher currents than this can cause problems of heating inside the capillary which can broaden peaks resulting in loss of resolution.

Selected from "English for Chemistry and Chemical Engineering, by Zhang Yuping etc., Chemical Industry Press, 2007,185-189"

New Words

1. chromatography [ˌkrəumə'tɔɡrəfi] *n.* 色 | 谱法，层析法，层离法，色层分析法

2. capillary electrophoresis 毛电管电泳
3. chalk [tʃɔːk] n. 白垩，方解石
4. column [ˈkɔləm] n. 柱，支柱，圆柱
5. pigments [ˈpigmənt] n. 色素，颜料
6. stationary [ˈsteiʃ(ə)nəri] a. 固定的，不动的，n. 固定
7. analyte [ˈænəlait] n. 被分析物
8. immobile [iˈməubail] a. 不能移动的，固定的，静止的
9. pack [pæk] vt. 填充，包装，压紧
10. manipulate [məˈnipjuleit] vt. 熟练的操作，使用机器的，利用
11. versatility [ˌvɜːsəˈtiliti] n. 多功能性
12. injector [inˈdʒektə] n. 注射器
13. filter [ˈfiltə] n. 过滤器，滤光器，筛选；vt. 过滤，渗透；vi. 滤过，渗入
14. regulator [ˈregjuleitə] n. 调节器，稳定器
15. sparging [ˈspɑːdʒiŋ] n. 起泡，鼓泡，喷射
16. propel [prəˈpel] vt. 推进，驱使
17. seal [siːl] n. 封铅，封条 图章 密封；vt. 封，密封
18. valve [vælv] n. 阀
19. loop [luːp] n. 环，回路，循环；vt. 使成环，以环连接；vi. 打环
20. reproducible [ˌriːprəˈdjuːsəbl] a. 可再生的，可复写的
21. sacrificial [ˌsækriˈfiʃəl] a. 牺牲的
22. foul [faul] vt. 弄脏，淤塞；vi. 腐烂，缠结；a. 淤塞的
23. digitize [ˈdidʒitaiz] vt. 将数字化，数字化处理
24. recorder [riˈkɔːdə] n. 记录仪，记录员，录音机
25. install [inˈstɔːl] vt. 安装，安置
26. eluent [ˈeljuənt] n. 洗提液
27. retard [riˈtɑːd] vt. 延迟，使减速，阻止，阻碍
28. retention [riˈtenʃən] n. 保留，保持
29. porosity [pɔːˈrɔsiti] n. 多孔性，有孔性，空隙率
30. non-destructive a. 无破坏性的，无害的
31. labile [ˈleibail] a. 易发生变化的，易分解的，不稳定的
32. derivatization n. 衍生法
33. pictogram [ˈpiktəˌgræm] n. 皮克，微微克
34. visualization [ˌviʒuəlaiˈzeiʃən] n. 使看得见的
35. dissipated [ˈdisipeitid] a. 散逸的，浪费的
36. buffer [ˈbʌfə] n. 缓冲液，缓冲器
37. capillary zone electrophoresis 毛细管区带电泳
38. free solution CE 自由溶液毛细管电泳
39. micelle [miˈsel] n. 胶束，胶囊
40. micellar electrokinetic capillary chromatography 胶束电动毛细管色谱
41. sieving [ˈsiviŋ] n. 筛选，筛分法
42. capillary gel eletrophoresis 毛细管凝胶电泳
43. zwitterionic [ˌzwitəraiˈɔnik] a. 两性离子的
44. capillary isoelectric focusing 毛细管等电聚焦
45. electroosmotic [iˌlektrəuɔzˈməutik] a. 电渗的
46. capillary electrochromatography 毛细管电色谱

Phrases

dependent upon (depend on) 依靠的，依赖的，由……决定的，随……而定的
break down 毁掉，制服，压倒，停顿，倒塌，中止，垮掉，分解
interfere…with 干涉，干扰，妨碍
be removed to 移动
be interfaced with 与……连接

Affixes

re- 表示"又，再"之意，如：reproducible, reaccess, reable, refresh
pre- 表示"之前"如：precolumn, preadult preabsorption

Notes

1. Liquid Chromatography was first discovered in 1903 by M. S. Tswett, who used a chalk column to separate the pigments of green leaves. 为非限制性定语从句, who 引导的从句对前面的主句起进一步解释和说明的作用。参考译文:液相色谱最早是由 M. S. Tswett 发现的,他使用石灰柱分离了绿叶中的色素。

2. However, a sacrificial guard column is often included just prior to the analytical column to chemically remove components of the sample that would otherwise foul the main column. that 引导的定语从句修饰 components，prior to 的意思是 "在前,居先"。参考译文:通常，紧靠分析柱前还会有牺牲保护柱,其功能是用化学方法除去样品中可能对主分析柱造成危害的组分。

3. Assume that component A has the same interaction with the adsorbent surface as an eluent, and component B has strong excessive interaction. the same ⋯as 的意思是 "与⋯⋯的⋯⋯一样"。参考译文:假定 A 组分和洗涤液对吸附剂表面的亲和力相同,而 B 组分对吸附剂表面的亲和力却比其对洗液的亲和力大得多。

4. The method is popular because it is non-destructive and may be applied to thermally labile compounds(unlike GC);it is also a very sensitive technique since it incorporate a wide choice of detection methods. because 是并列连词,表示 "原因"。句中 the method 和三个 it 都是指 HPLC。参考译文：高效液相色谱因为不会破坏样品以及可用于检测热不稳定成分（与气相色谱不同）而被广泛使用；高效液相色谱也是非常灵敏的技术，因为它可以有非常多的检测方法。

5. With the use of post-column derivatization methods to improve selectivity and detection limits, HPLC can easily be extended to trace determination of compounds that do not usually provide adequate detector response。that 引导的定语从句用来修饰 compound。参考译文：有些痕量组分通常不能产生足够检测响应，如果采用柱后衍生法提高选择性和检测限，则高效液相色谱还可用于痕量组分的检测。

6.The variations include separation based on size and charge differences between analytes（termed Capillary Zone Electrophoresis，CZE，or Free Solution CE，FSCE），separation of neutral compounds using surfactant micelles（Micellar electrokinetic capillary chromatography，MECC or sometimes referred to as MEKC），sieving of solutes through a gel network（Capillary Gel Electrophoresis，CGE），and separation of zwitterionic solutes within a pH gradient（Capillary Isoelectric Focusing，CIEF）. 参考译文：具体包括基于被分析物的大小和电荷差别而进行的分离（毛细管区带电泳，CZE，或者自由溶液电泳，FSCE），利用表面活性剂胶束（胶束电动毛细血管色谱，MECC 或者 MEKC）分离电中性化合物，通过凝胶网络来筛分溶质（毛细管凝胶电泳，CGE），以及利用 pH 梯度（毛细管等电聚焦，CIEF）来分离两性溶质。

Lesson 21
Membranes for Separation Process

Membranes for industrial separation processes can be classified into the following groups according to the driving force that causes the flow of the permeant through the membrane.

1) Pressure difference across the membrane is the driving force.
   Reverse osmosis
   Ultrafiltration
   Microfiltration
   Membrane gas and vapor separation
   Pervaporation
2) Temperature difference across the membrane is the driving force.
   Membrane distillation
3) Concentration difference across the membrane is the driving force.
   Dialysis
   Membrane extraction
4) Electric potential difference across the membrane is the driving force.
   Electrodialysis

Reverse osmosis is a process to separate solute and solvent components in the solution. Although the solvent is usually water, it is not necessarily restricted to water. The pore radius of the membrane is less than 1 nanometer (nm) (1m = 10^9nm). While solvent water molecules, whose radius is about one tenth of 1 nm, can pass through the membrane freely, electrolyte solutes, such as sodium chloride and organic solutes that contain more than one hydrophilic functional group in the molecule (sucrose, for example), cannot pass through the membrane. These solutes are either rejected from the membrane surface, or they are more strongly attracted to the solvent water phase than to the membrane surface. The preferential sorption of water molecules at the solvent-membrane interface, which is caused by the interaction force working between the membrane-solvent-solute, is therefore responsible for the separation. Polymeric materials such as cellulose acetate and aromatic polyamide are typically used for the preparation of reverse osmosis membranes. A thin, dense layer responsible for the separation, and therefore often called an active surface layer, is supported by a porous sublayer that provides the membrane with sufficient mechanical strength. The entire thickness of the membrane is about 0.1 mm, while the thickness of the active surface layer is only 30 to 100 nm.

When a membrane is placed between pure water and an aqueous sodium chloride solution, water flows from the chamber filled with pure water to that filled with the sodium chloride solution, whereas sodium chloride does not flow. As water flows into the sodium chloride solution chamber, the water level of the solution increases until the flow of pure water stops at the steady state. The difference between the water level of the sodium chloride solution and that of pure water at the steady state, when converted to hydrostatic pressure, is called osmotic pressure. When a pressure higher than the osmotic pressure is applied to the sodium chloride solution, the flow of pure water is reversed; the flow from the sodium chloride solution to the pure water begins to occur. There is no flow of sodium chloride through the membrane. As a result, pure water can be obtained from the sodium chloride solution. The above separation process is called reverse osmosis.

The most successful application of the reverse osmosis process is in the production of drinking water from seawater. This process is known as seawater desalination and is currently producing millions of gallons of potable water daily in the Middle East. Fishing boats, ocean liners, and submarines also carry reverse osmosis units to obtain potable water from the sea. In brackish water where the content of sodium chloride is much less than in seawater, lower osmotic pressures should be overcome for desalination. The reverse osmosis process is also being used to produce ultrapure water for the manufacture of semiconductors.

Ultrafiltration is a process based on the same principle as that of reverse osmosis. The main difference between reverse osmosis and ultraflltration is that ultrafiltratinn membranes have larger pore sizes than reverse osmosis membranes, ranging from 1 to 100 nm. Ultrafiltration membranes are used for the separation and concentration of macromolecules and colloidal particles. Osmotic pressures of macromolecules are much smaller than those of small solute molecules, and therefore operating pressures applied in the ultrafiltration process are usually much lower than those applied in the reverse osmosis process. Membranes having pore sizes between those for reverse osmosis and ultrafiltration membranes are sometimes called nanofiltration membranes. The size of the solute molecules that are separated from water, and the range of operating pressures, are also between those for reverse osmosis and ultrafiltration. Ultrafiltration membranes are prepared from polymeric materials such as polysulfone, polyethersulfone, polyacrylonitrile, and cellulosic polymers. Inorganic materials such as alumina can also be used for ultrafiltration membranes.

Typical applications of ultrafiltration processes are the treatment of electroplating rinse water, the treatment of cheese whey, and the treatment of wastewater from the pulp and paper industry.

The pore sizes of microfiltration membranes are even larger than those of ultrafiltration membranes and range from 0.1μm (100 nm) to several μm. The sizes of the particles separated by microfiltration membranes are therefore even larger than those separated by ultrafiltration membranes. However, the separation mechanism is not a simple sieve mechanism whereby the particles whose sizes are smaller than the pore size flow freely through the pore while the particles that are larger than the pore size are stopped completely. In many cases the particles to be separated are adsorbed onto the surface of the pore, resulting in a significant reduction in the pore size. Particles can also be deposited on top of the membrane, forming a cake-like secondary filter layer. Therefore, the sizes of the particles that can be separated by microfiltration membranes are often much smaller than those of the pores in the "uncontaminated" membranes. Several methods are used to prepare microfiltration membranes. One of the methods is to sinter small particles made of metals, ceramics, and plastics. The spaces formed between the particles become the pores through which materials can be transported. A second method is to stretch a polymeric film. When a polyethylene film is stretched, part of the film becomes opaque. Pores are found in this part by electron micrographic observation. Another method is to irradiate a plastic film, such as a polycarbonate film, with an electron beam. Pores are formed when sections hit by the electron beam are chemically etched with a strong alkaline solution. The phase-inversion technique, in which the polymeric solution is cast onto a film and then solidified by immersing the film in a nonsolvent gelation bath, is also applied to prepare microfiltration membranes.

Microfiltration membranes are used for the removal of microorganisms from the fermentation product. For example, various antibiotics are produced by the function of microorganisms. Microfiltration membranes are used to separate the microorganisms from the product antibiotics. Microfiltration membranes are also used to remove yeast from alcoholic beverages. For example, in the process of producing draught beer, yeast is removed by membrane filtration. Recently, a cartridge for cleaning tap water was developed. Microfiltration hollow fibers made of polyethylene are combined in the cartridge with calcium carbonate and activated carbon columns. Small organic molecules, such as halogenated hydrocarbons, are removed by adsorption to activated carbon. Mineral and carbon dioxide contents in water are increased while passing through a calcium carbonate layer. Finally, microorganisms, molds, and other turbid materials are removed by filtration with microfiltration hollow fibers.

Selected from "English for Chemical Engineers, by Ma Zhengfei etc., Southeast University Prss, 2006, 147-149."

Exercises

1. Put the following into Chinese:

    hydrostatic pressure   osmotic pressure   pervaporation   polyamide
    dialysis   sublayer   electrodialysis desalination
    membrane extraction   submarine   brackish   opaque
    macromolecule   ceramic   fermentation   implementation

2. Put the following into English:

    超滤   微滤   反渗透   制药的
    污垢   决定性的   合成   组件
    突破   混合的   亲水的   疏水的

3. Comprehension and toward interpretation

    a. What are the main applications of membrane technology?
    b. For membrane technology to replace existing technology, what should do?
    c. What are the advantages of membrane technology?
    d. There are not much membrane applications in industry, why?

Reading Material

Membranes in Chemical Processing

The development of membrane technology will strongly influence the way industry is looking to separation processes in the future. In some fields membranes are already "proven technology" and incorporated in various production lines or purification processes. This is mainly within the food and dairy industries, water purification and treatment of liquid effluent streams. Some membranes units for pharmaceutical and medical applications are common, but large units in the chemical process industry are still rather rare. The current situation is due to several factors of both technical and economical kind. The lifetime of membrane material is crucial for the integration of membrane modules in industrial process streams which will usually transport fairly large volumes of gas and liquids at pressures and temperatures where the durability of the membrane materials over time are not yet fully exploited[1]. If the membranes have to be replaced too often, the solution may be too expensive, or if the membrane is easily damaged it may be dangerous as well. First of all the basic question must be asked: Will the membrane solution as well or better than the existing technology? This question has to be answered on the basis of product quality, energy consumption and environmental issues. Secondly: are the costs of this new technology currently at a level which will make the implementation attractive? Not until this achieved for the various applications, will this technology have a significant technical breakthrough.

More than in any other field of separation, membrane technology demands that basic research within material science is coupled to the understanding of problems related to the specific industrial process where the membrane module is to integrated[2]. Too often research stops in a laboratory with experiments carried out under ideal conditions and over too short a time period, and results as "interesting and promising" for a certain membrane. A stronger involvement of industry is often necessary in order to develop the membrane to commercial level, and thus promote the incorporation of membrane modules in a process together with other unit operations[3]. The argument for doing so is the obvious advantages this technology offer: cleaner and simpler solutions, less chemical additives, lower energy consumption.

The demand on industry for better environmental solutions and cleaner technology is thus pushing membrane technology into the spotlight[4]. End-of-pipe solutions for purification of effluent streams will to a larger extent be substituted by closed systems with integrated process solutions in the future, resulting in simpler and cleaner industrial processes. Membranes are the solution nature itself uses for separation and purification—the technical challenge is to develop just as efficient synthetic membranes for the separation of industrial process streams.

Membrane technology is presently being introduced into a wide variety of applications, and there is clearly a very positive trend for the development of industrial membrane applications. For applications related to water and wastewater treatment, the focus is partly on how to reduce fouling and improve water flux. Improvement of the membrane materials may be done by using various types of surface treatment. However, the trend shows that new applications in more harsh industrial environments often demand other types than standard hydrophilic materials for treatment effluent streams. For applications related to gas separation and purification, membrane technology has up until now been common only within a few areas. More than within liquid separation, the membrane material plays an active part in the transport of the gases through the membrane. Evaluating this trend, conventional polymeric materials appear to have only marginal room for improvements[5], although reactive surface flow modifications may be useful in some cases. With the development of new materials for facilitated transport however, carbon molecular sieve[6] membranes, catalytic membrane reactors and mixed matrix materials, the potential for application of gas separation membranes is very promising.

Process solutions using integrated hybrid membrane systems will often be the best one to a specific industrial separation problem; hence good engineering combined with in-depth knowledge of the membranes is important, indeed critical, for acceptance of the technology within chemical industrial processes[7].

Selected from "English for Chemical Engineers, by Ma Zhenfei etc., Southeast University Press, 2006, 144-146"

New Words

1. membrane ['membrein]   n. 膜，薄膜
2. proven ['pru:vn]   a. 验证的
3. pharmaceutical [fɑ:mə'sju:tikəl]   a. 药的，制药的
4. crucial ['kru:ʃəl]   a. 决定性的，重要的
5. module ['mɔdju:l]   n. 组件
6. implementation [implimen'teiʃən]   n. 实现，实施，履行
7. breakthrough ['breikθru:]   n. 穿过，透过，突破
8. argument ['ɑ:gjumənt]   n. 争论，论证
9. synthetic [sin'θetik]   a. 合成的
10. fouling [fauliŋ]   n. 污垢
11. hydrophilic [haidrə'filik]   a. 亲水的
12. hybrid ['haibrid]   a. 杂交的，混合的

Notes

1. The lifetime of membrane material is crucial for the integration of membrane modules in industrial process streams which will usually transport fairly large volumes of gas and liquids at pressures and temperatures where the durability of the membrane materials over time are not yet fully exploited. 参考译文：将膜组件结合到工业过程流程中，膜材料的寿命是关键因素，因通常在工业过程流程中，需输送大量具有一定压力和温度的气体和液体，此种情况下的膜材料寿命还没有进行充分研究。

2. More than in any other field of separation, membrane technology demands that basic research within material science is coupled to the understanding of problems related to the specific industrial process where the membrane module is to integrated. 参考译文：和其他任何领域的分离相比，膜技术更需要材料科学中的基础研究同了解耦合有膜组件的特殊工业过程相关联的问题相结合。

3. A stronger involvement of industry is often necessary in order to develop the membrane to commercial level, and thus promote the incorporation of membrane modules in a process together with other unit operations. 参考译文：为了将膜应用于实际过程，同工业界的紧密合作常是必需的，这样可促进过程中膜组件同其他单元操作的结合。

4. push something into the spotlight   使……成为中心。

5. have only marginal room for improvements   只有很小的改进余地。

6. carbon molecular sieve   碳分子筛。

7. Process solutions using integrated hybrid membrane systems will often be the best one to a specific industrial separation problem; hence good engineering combined with in-depth knowledge of the membranes is important, indeed critical, for acceptance of the technology within chemical industrial process. 参考译文：对特殊的工业分离问题，用混杂的有膜系统的过程方法常会是最好的方法。因此，精深的膜知识结合良好的工程对在化学工业过程中使用膜技术是重要的，可以说是关键的。

Lesson 22
New Technologies in Unit Operations

While technical advances and efficiency improvements in specific unit operations are occurring all tile time, the big story is the hybridization of processes. Combining individual unit operations, such as reaction, separation, and heat exchange, into larger, concurrent operations will be major trend in upcoming years[1]. Technologies such as reactive distillation, catalytic membranes, and phase-transfer catalysis[2] all represent examples of hybridized processes where reaction and separation are combined. A major advantage of these combined operations is a huge reduction in capital expenditure[3] — typically to one-tenth to one-fifth of the investment for a traditional setup[4]. In practice, at least at first while confidence in the performance and reliability of the combined operations builds[5], companies usually will run the newer, hybridized operations in parallel with older processes for the same product.

In addition to their significant reduction in capital outlay, combined reaction/separation processes offer two other major advantages: reduction of unwanted reaction byproducts, and improvement of yields for reactions with low equilibrium constants.

With environmental issues driving so many of the changes in the CPI[6], technologies that help reduce reaction byproducts clearly are of increasing value. With combined reaction/separation, highly reactive feedstock does not have a chance to react with products because the products are separated out immediately after formation. This also reduces waste of feedstock and product. Immediate removal of reaction products also will enable low-equilibrium-constant reactions that are not commercially feasible under normal conditions to proceed much further. Indeed, reactions that go to only a few percent conversion can be forced to 100% conversion through combined reaction/separation. Several processes in this category are currently under development, and large commercial operations are currently being designed and built for them.

A couple of examples of recent applications of these hybrid processes illustrate the benefits of this new approach to unit operations. Reactive distillation is proving very useful for the production of ethers, such as methyl tert-butyl ether (MTBE), tert-amyl methyl ether (TAME), and ethyl tert-butyl ether (ETBE)[7], which increasingly are employed to boost oxygenate content in gasoline[8]. Here, the preferred temperature range for the catalyst is the same as that for the distillation of ethers from reactants and inerts. Thus, reactive distillation provides energy savings, as well.

Catalytic membranes are a newer example of combined reaction/separation processes. In this scheme, the catalyst material also acts as a sieving system to separate reaction products as they are formed. A key advantage of membrane separation processes is their energy efficiency; they also particularly suit heat-sensitive material[9], such as pharmaceuticals and foodstuffs. Dense catalytic membranes, which separate, for example, on the basis of gas diffusivity, are closer to commercial deployment than are porous membranes, which separate on the basis of molecular size. Robust processes have not yet emerged, however, for synthesizing defect-free porous membranes[10] with appropriate composition and pore size range for the variety of catalytic reactions and gas adsorption applications that exist in the CPI.

A second and newer trend in unit operations is the advent of the minireactor. The idea of "desktop chemical manufacturing" takes the large-scale, continuous commodity plant and scales it down for the manufacture of speciality chemicals such as pharmaceuticals[11]. Instead of batch processing different chemicals one after another in the same equipment[12], a plant might consist of several small systems running continuously throughout the year. The advantages include better consistency of product, simpler scheduling and monitoring, and standardized small-scale equipment. Although a consensus has not yet emerged about the overall economic value of the minireactor approach, early industrial trials have shown that there may be some significant advantages.

This minireaetor approach has even been extended to the idea of the microreactor. University and industrial researchers have developed a prototype microfluidic reactor[13] that measures no more than 2 square cm. Proposed reaction systems for the unit include hydrogen and methane oxidation, ethylene epoxidation, and phosgene synthesis[14]. Once optimized, such a microreactor then can be scaled up to commercial proportions through simple replication and arrangement of the individual units. The potential advantages of such a system include better process safety relative to macroscale reactors[15], and improved ability to integrate control, sensor, and reactor functionality[16].

Selected from "English for Chemical Engineers, by Ma Zhengfei etc., Southeast University Press, 2006, 17-20"

New words

1. hybridization [haibridaiˈzeiʃən] n. 杂交，杂化
2. concurrent [kənˈkʌrənt] a. 并流的，顺流的
3. byproduct [ˈbaiprɔdʌkt] n. 副产物
4. feedstock [fiːdˈstɔk] n. 原料
5. product [ˈprɔdʌkt] n. 产物
6. conversion [kənˈvəːʃən] n. 转化，转换
7. ether [ˈiːθə] n. 醚
8. inert [iˈnəːt] n. 惰性组分
9. sieving [siviŋ] n. 筛分
10. diffusivity [difjuːˈsiviti] n. 扩散性，扩散系数
11. deployment [ˈdiplɔimənt] n. 使用，利用，推广应用
12. robust [rəuˈbʌst] a. 加强的，增强的，健全的
13. adsorption [ædˈsɔːpʃən] n. 吸附
14. consensus [kənˈsensəs] n. 一致
15. trial [ˈtraiəl] n. 试验，审判
16. approach [əˈprəutʃ] n. 途径，方法，手段
17. replication [repliˈkeiʃən] n. 重复实验

Notes

1. Combining individual unit operations, such as reaction, separation, and heat exchange, into large, concurrent operations will be major trend in upcoming years. 参考译文：未来若干年中，综合各单元操作，例如反应、分离和热交换，构成较大的同时操作将成为主要趋势。

2. Technologies such as reactive distillation, catalytic membranes, and phase-transfer catalysis 参考译文：例如反应精馏、催化膜和相传递催化等技术。

3. capital expenditure 基建投资。

4. traditional setup 传统装置。

5. while confidence in the performance and reliability of the combined operations builds. 参考译文：在建立对综合操作的性能和可靠性的信心时。

6. With environmental issues driving so many of the changes in the CPI. 参考译文：由于环境问题促使化学制造工业出现许多变化。

7. such as methyl tert-butyl ether (MTBE),tert-amyl methyl ether (TAME),and ethyl tert-butyl ether (ETBE)。参考译文：例如甲基叔丁基醚（METB）、叔戊基甲基醚（TAME）和乙基叔丁基醚（ETBE）。

8. to boost oxygenate content in gasoline. 强化汽油充氧量。

9. heat-sensitive material 热敏材料。

10. defect-free porous membranes 无缺陷多孔膜。

11. The idea of "desktop chemical manufacturing" takes the large-scale, continuous commodity plant and scales it down for the manufacture of specialty chemicals such as pharmaceuticals. 参考译文："化学制造工作台"利用了大规模、连续化商品生产工厂的概念，并缩小制造如药品这样的特殊化学品的规模。

12. Instead of batch processing different chemicals one after another in the same equipment 参考译文：不是在同一设备中相继间歇生产不同的化学品。

13. prototype microfluidic reactor 原型微流反应器。

14. hydrogen and methane oxidation, ethylene epoxidation, and phosgene synthesis. 参考译文：氢和甲烷氧化、乙烯环氧化和光气合成。

15. macroscale reactors 大尺寸反应器。

16. improved ability to integrate control, sensor, and reactor functionality. 参考译文：改进集成控制、传感器和反应器功能的能力。

Exercises

1. Put the following into Chinese:
   conversion   inert   trial   replication
   ether   sieving   approach   hybridization
   robust   magnetic   biomaterial   superconducting

2. Put the following into English:
   并流   扩散系数   催化膜   副产物
   吸附   微流反应器   原料   热敏材料
   大尺寸反应器   产物   反应精馏   装置

3. Comprehension and toward interpretation
   a. In which areas do new technologies appear?
   b. Environmental issues drive the changes in chemical processing industry. Why?
   c. What is the function of MTBE in gasoline?
   d. What are the benefits of new technologies?

Reading Material

What Is Materials Science and Engineering?

Where Did Materials Science and Engineering (MSE) Come from?

Although materials and processes have fueled technological progress for thousands of years, the field of MSE per se did not exist before the 1960s. It is, therefore, a relatively young discipline in comparison with physics, chemistry, and related engineering fields. MSE became a single discipline through the evolution and coalescence of three materials—specific fields— metallurgy, ceramics, and polymer science. Although many other disciplines—for example, physics, geology, electronics, optics, chemistry, and biology—continue to bear on MSE and have made indispensable contributions to its development as a formal discipline, these three material-based fields remain at the heart of MSE.

In the early days of MSE as an academic discipline and a subject of R&D endeavor, practitioners came mostly from physics, chemistry, engineering, metallurgy, the earth sciences, and mathematics. With the growth of biomaterials, medical practitioners, biologists, biochemists, and biophysicists have joined in. Increasingly, as the field emerged in its own fight, MSE practitioners were trained in materials departments established at engineering or physical sciences schools in universities home and abroad. Nevertheless, despite 50 years of developing, maturing, and gaining broad acceptance agreeing on an all-encompassing definition for MSE as a discipline remains a challenge. Tile origins and nature of MSE remain varied and interwoven, and any definition of the field must reflect the richness and diversity of all the activity related to "materials".

What Is a Material?

A good place to start defining MSE is to consider what a material is. A simple definition would be that a material is the stuff from which an article, fabric, or structure is made. This definition, however attractive because of its simplicity, does not reflect the full diversity of the study of materials. Because most articles, fabrics, or structures are considered to be solids, how would the study of liquids and gases fit into such a definition of materials? It would not, yet the study of liquids and gases is of central importance to many areas of MSE, such as materials processing, understanding the structure of many biological systems, investigating colloidal systems, and studying liquid crystals. A more thorough definition might be this: Matter is a "material" when that form of matter has structural, optical, magnetic, or electrical use.

What Do Materials Scientists and Engineers Do?

Further insight into what is meant by MSE can he gained by considering the kinds of things materials scientists study and the material-related knowledge and skills they need to do so.

Structure. Electronic, atomic, bonding, crystalline, amorphous, and multiphase structure on the nano-, micro-, meso-, and macroscales.

Characterization of Composition and Microstructure. Spectroscopy, optical and electron microscopy, electron and X-ray diffraction, scanning probe techniques, thermal analysis, and some aspects of traditional chemical analysis.

Phase Equilibria and Phase Transformations. Thermodynamic and kinetic aspects.

Mechanical Behavior. Elastic and plastic deformation and fracture; strengthening, toughening, and stiffening mechanisms; mechanical test methods; and continuum mechanics.

Functional Behavior. Semiconducting, dielectric, optical, conducting, and magnetic materials and materials that interact with or draw inspiration from biological systems.

Processing and Manufacture. Processing and synthesis of materials via gaseous, liquid, colloidal, powder, solid state, and deposition techniques; joining and fabrication methods; surface treatment; heat and mass transfer; and fluid mechanics.

Degradation and Durability of Materials. Effect of liquid and gaseous environments on the performance of different material types, wear of material, and biodegradation.

Materials Selection. Consideration of all material types, including material processing method, life-cycle analysis and product costs; selection criteria for materials and production processes.

Design with Materials. The selection of appropriate compositions; choice of and use of processing and manufacture to achieve the required microstructure, structural, and functional properties in a product according to agreed specifications.

It has been suggested that these activities can be summarized by considering that materials scientists and engineers investigate the function of a material in an existing application and discover applications for it through the characterization of its structure and properties, and through the understanding of its production. The materials tetrahedron (see Fig.22-1) is often used to illustrate the four aspects of MSE: (1) composition and microstructure, (2) properties, (3) synthesis and processing, and (4) performance. Each is intellectually rich and challenging, and together they help define the discipline of MSE in its own right.

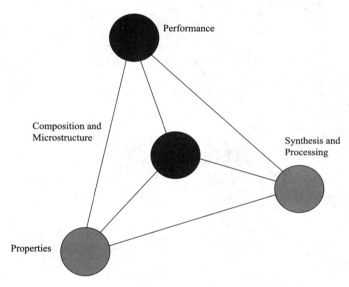

Fig. 22-1   The MSE tetrahedron reflects the four fundamental aspects of a material—performance, properties, synthesis and processing, and composition and microstructure. While the four aspects can be treated individually as disciplines, a comprehensive approach to the materials tetrahedron is usually required for complex problems.

Because of the enormous breadth of the field, MSE R&D must be divided into subfields. In its 1993 and 1995 reports, the National Science and Technology Council (NSTC) listed the subfields of materials research and engineering as biomaterials, ceramics, composites, electronic materials, magnetic materials, metals, optical-photonic materials, polymers, and superconducting materials. The committee that carried out the MSE benchmarking exercise added catalysts to the NSTC list and combined electronic materials and optical-photonic materials research into one category. Since publication of the benchmarking report, the investigation of nanomaterials has blossomed into a vibrant, crosscutting area of materials research, so for the purposes of this report, the subfield of nanomaterials has been added to the list.

MSE at the beginning of the 21st century, therefore, consists of the following subfields:
Biomaterials
Ceramics

- Composites
- Magnetic materials
- Metals
- Electronic and optical-photonic materials
- Superconducting materials
- Polymers
- Catalysts
- Nanomaterials

In summary, MSE involves the generation and application of knowledge that relates the composition, structure, and processing of materials to their properties and uses.

Selected from "English for Chemical Engineers, by Ma Zhengfei etc., Southeast University Press, 2006, 206-208"

Lesson 23
Structure and Nomenclature of Hydrocarbons

The Saturated Hydrocarbons, or Alkanes

Compounds that contain only carbon and hydrogen are known as **hydrocarbons.** Those that contain as many hydrogen atoms as possible are said to be saturated. The saturated hydrocarbons are also known as alkanes.

The simplest alkane is methane: CH_4. The Lewis structure of methane can be generated by combining the four electrons in the valence shell of a neutral carbon atom with four hydrogen atoms to form a compound in which the carbon atom shares a total of eight valence electrons with the four hydrogen atoms.

Methane is an example of a general rule that carbon is **tetravalent**; it forms a total of four bonds in almost all of its compounds. To minimize the repulsion between pairs of electrons in the four **C—H** bonds, the geometry around the carbon atom is tetrahedral, as shown in Fig. 23-1.

The names, formulas, and physical properties for a variety of alkanes with the generic formula C_nH_{2n+2} are given in Table 23-1. The boiling points of the alkanes gradually increase with the molecular weight of these compounds, At room temperature, the lighter alkanes are gases; the midweight alkanes are liquids; and the heavier alkanes are solids, or tars.

Fig. 23-1 Geometry of methane

The alkanes in the table below are all straight-chain hydrocarbons, in which the carbon atoms form a chain that runs from one end of the molecule to the other. The generic formula for these compounds can be understood by assuming that they contain chains of CH_2 groups with an additional hydrogen atom capping either end of the chain[1]. Thus, for every n carbon atoms there must be $2n+2$ hydrogen atoms: C_nH_{2n+2}.

Table 23-1 The Saturated Hydrocarbons, or Alkanes

Name	Molecular Formula	Melting Point/℃	Boiling Point/℃	State at 25 ℃
methane	CH_4	−182.5	−164	gas
ethane	C_2H_6	−183.3	−88.6	gas
propane	C_3H_8	−189.7	−42.1	gas
butane	C_4H_{10}	−138.4	−0.5	gas
pentane	C_5H_{12}	−129.7	36.1	liquid

Name	Molecular Formula	Melting Point/℃	Boiling Point/℃	State at 25 ℃
hexane	C_6H_{14}	−95	68.9	liquid
heptane	C_7H_{16}	−90.6	98.4	liquid
octane	C_8H_{18}	−56.8	124.7	liquid
nonane	C_9H_{20}	−51	150.8	liquid
decane	$C_{10}H_{22}$	−29.7	174.1	liquid
undecane	$C_{11}H_{24}$	−24.6	195.9	liquid
dodecane	$C_{12}H_{26}$	−9.6	216.3	liquid
icosane	$C_{20}H_{42}$	36.8	343	solid
triacontane	$C_{30}H_{62}$	65.8	449.7	solid

Because two points define a line, the carbon skeleton of the ethane molecule is linear, as shown in Fig. 23-2.

Fig.23-2 Carbon skeleton of ethane molecule

Because the bond angle in a tetrahedron is 109.5, alkanes molecules that contain three or four carbon atoms can no longer be thought of as "linear", as shown in Fig. 23-3.

Fig. 23-3 Carbon skeleton of propane and butane

In addition to the straight-chain examples considered so far, alkanes also form branched structures. The smallest hydrocarbon in which a branch can occur has four carbon atoms. This compound has the same formula as butane (C_4H_{10}), but a different structure. Compounds with the same formula and different structures are known as isomers (from the Greek isos, "equal", and meros, "parts"). When it was first discovered, the branched isomer with the formula C_4H_{10} was therefore given the name isobutane.

$$CH_3-CH(CH_3)-CH_3 \quad \text{Isobutane}$$

The best way to understand the difference between the structures of butane and isobutane is to compare the ball-and-stick models of these compounds shown in Fig.23-4.

Butane and isobutane are called constitutional isomers because they literally differ in their constitution. One contains two CH_3 groups and two CH_2 groups; the other contains three CH_3 groups and one CH group.

Fig.23-4   Comparison of ball-and stick models of butane and isobutane

There are three constitutional isomers of pentane, C_5H_{12}. The first is "normal" pentane, or n-pentane.

$$CH_3-CH_2-CH_2-CH_2-CH_3 \quad \text{n-Pentane}$$

A branched isomer is also possible, which was originally named isopentane. When a more highly branched isomer was discovered, it was named neopentane (the new isomer of pentane).

$$\begin{array}{c}CH_3\\|\\CH_3-CH-CH_2-CH_3\end{array} \quad \text{Isopentane}$$

$$\begin{array}{c}CH_3\\|\\CH_3-C-CH_3\\|\\CH_3\end{array} \quad \text{Neopentane}$$

Ball-and-stick models of the three isomers of pentane are shown in Fig. 23-5.

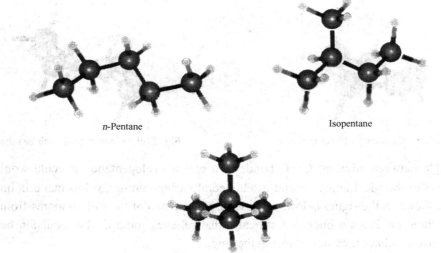

Fig. 23-5   Ball-and-stick models of three isomers of pentane

There are two constitutional isomers with the formula C_4H_{10}, three isomers of C_5H_{12}, and five isomers of C_6H_{14}. The number of isomers of a compound increases rapidly with additional carbon atoms. There are over 4 billion isomers for $C_{30}H_{62}$, for example.

The Cycloalkanes

If the carbon chain that forms the backbone of a straight-chain hydrocarbon is long enough, we can envision the two ends coming together to form a cycloalkane. One hydrogen atom has to be removed from each end of the hydrocarbon chain to form the C—C bond that closes the ring. Cycloalkanes therefore have two less hydrogen atoms than the parent alkane and a generic formula of C_nH_{2n}.

The smallest alkane that can form a ring is cyclopropane, C_3H_6, in which the three carbon atoms lie in the same plane. The angle between adjacent C—C bonds is only 60, which is very much smaller than the 109.5 angle in a tetrahedron, as shown in Fig. 23-6.

Cyclopropane is therefore susceptible to chemical reactions that can open up the three-membered ring.

Any attempt to force the four carbons that form a cyclobutane ring into a plane of atoms would produce the structure shown in Fig. 23-7, in which the angle between adjacent C—C bonds would be 90.

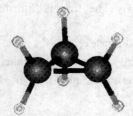 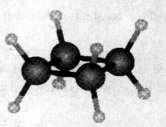

Fig. 23-6　Geometry of cyclopropane　　　Fig. 23-7　Geometry of cyclobutane

One of the four carbon atoms in the cyclobutane ring is therefore displaced from the plane of the other three to form a "puckered" structure that is vaguely reminiscent of the wings of a butterfly[2] (see Fig. 23-8).

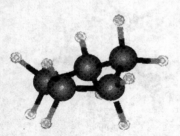 

Fig. 23-8　Geometry of cyclopentane　　　Fig. 23-9　Geometry of cyclohexane

The angle between adjacent C—C bonds in a planar cyclopentane molecule would be 108, which is close to the ideal angle around a tetrahedral carbon atom. Cyclopentane is not a planar molecule, as shown in the figure below, because displacing two of the carbon atoms from the plane of the other three produces a puckered structure that relieves some of the repulsion between the hydrogen atoms on adjacent carbon atoms in the ring.

By the time we get to the six-membered ring in cyclohexane, a puckered structure can be formed by displacing a pair of carbon atoms at either end of the ring from the plane of the other four members of the ring. One of these carbon atoms is tilted up, out of the ring, whereas the other is tilted down to form the "chair" structure shown in Fig. 23-9[3].

The Nomenclature of Alkanes

Common names such as pentane, isopentane, and neopentane are sufficient to differentiate between the three isomers with the formula C_5H_{12}. They become less useful, however, as the size of the hydrocarbon chain increases.

The International Union of Pure and Applied Chemistry (IUPAC) has developed a systematic approach to naming alkanes and cycloalkanes based on the following steps.

Find the longest continuous chain of carbon atoms in the skeleton structure. Name the compound as a derivative of the alkane with this number of carbon atoms. The following compound, for example, is a derivative of pentane because the longest chain contains five carbon atoms.

$$\begin{array}{c} H \\ | \\ H-C-H \\ H \quad | \quad H \quad H \quad H \\ | \quad | \quad | \quad | \quad | \\ H-C-C-C-C-C-H \\ | \quad | \quad | \quad | \quad | \\ H \quad H \quad H \quad H \quad H \end{array}$$

Name the substituents on the chain. Substituents derived from alkanes are named by replacing the -ane ending with -yl. This compound contains a methyl (CH_3-) substituent.

$$\begin{array}{c} H \\ | \\ H-C-H \\ H \quad | \quad H \quad H \quad H \\ | \quad | \quad | \quad | \quad | \\ H-C-C-C-C-C-H \\ | \quad | \quad | \quad | \quad | \\ H \quad H \quad H \quad H \quad H \end{array}$$

Number the chain starting at the end nearest the first substituent and specify the carbon aroms on which the substituents are located. Use the lowest possible numbers. This compound, for example, is 2-methylpentane, not 4-methypentane.

Use the prefixes di-, tri-, and tetra- to describe substituents that are found two, three, or four times on the same chain of carbon atoms.

$$\begin{array}{c} H \\ | \\ H-C-H \\ H \quad | \quad H \quad H \quad H \\ | \quad | \quad | \quad | \quad | \\ \overset{1}{H-C}-\overset{2}{C}-\overset{3}{C}-\overset{4}{C}-\overset{5}{C}-H \\ | \quad | \quad | \quad | \quad | \\ H \quad H \quad H \quad H \quad H \end{array}$$

Arrange the names of the substituents in alphabetical order.

The Unsaturated Hydrocarbons: Alkenes and Alkynes

Carbon not only forms the strong C—C single bonds found in alkanes, it also forms strong C=C double bonds. Compounds that contain C=C double bonds were once known as olefins (literally, "to make an oil") because they were hard to crystallize. (They tend to remain oily liquids when cooled.) These compounds are now called alkenes. The simplest alkene has the formula C_2H_4 and the following Lewis structure.

$$\begin{array}{c} H \quad \quad \quad H \\ \diagdown \quad \quad \diagup \\ C=C \\ \diagup \quad \quad \diagdown \\ H \quad \quad \quad H \end{array}$$

The relationship between alkanes and alkenes can be understood by thinking about the following hypothetical reaction. We start by breaking the bond in an H_2 molecule so that one of the electrons ends up on each of hydrogen atoms. We do the same thing to one of the bonds between the carbon atoms in an alkene. We then allow the unpaired electron on each hydrogen atom to interact with the unpaired electron on a carbon atom to form a new C—H bond.

$$\begin{array}{ccc} H-H & & \\ H \quad \quad H & H\cdot \quad H\cdot & H \quad H \\ \diagdown \quad \diagup & | \quad \quad | & | \quad \quad | \\ C=C & \longrightarrow \quad H-\dot{C}-\dot{C}-H & \longrightarrow \quad H-C-C-H \\ \diagup \quad \diagdown & | \quad \quad | & | \quad \quad | \\ H \quad \quad H & H \quad H & H \quad H \end{array}$$

Thus, in theory, we can transform an alkene into the parent alkane by adding an H_2 molecule across a C=C double bond. In practice, this reaction only occurs at high pressures in the presence of a suitable catalyst, such as piece of nickel metal.

Because an alkene can be thought of as a derivative of an alkane from which an H_2 molecule has been removed, the generic formula for an alkene with one C=C double bond is C_nH_{2n} [4].

Alkenes are examples of unsaturated hydrocarbons because they have fewer hydrogen atoms than the corresponding alkanes. They were once named by adding the suffix –ene to the name of the substituent that carried the same number of carbon atoms.

$$CH_2=CH_2 \quad \text{Ethylene}$$
$$CH_2=CH_2-CH_3 \quad \text{Propylene}$$

The IUPAC nomenclature for alkenes names these compounds as derivatives of the parent alkanes. The presence of the C=C double bond is indicated by changing the-ane ending on the name of the parent alkane to-ene.

$$CH_3-CH_3 \quad\quad CH_2=CH_2$$
$$\text{Ethane} \quad\quad\quad \text{Ethene}$$
$$CH_3-CH_2-CH_3 \quad\quad CH_3-CH=CH_2$$
$$\text{Propane} \quad\quad\quad\quad \text{Propene}$$

The location of the C=C double bond in the skeleton structure of the compound is indicate by specifying the number of carbon atom at which the C=C bond starts.

$$CH_2=CH-CH_2-CH_3 \quad\quad CH_3-CH=CH-CH_3$$
$$\text{1-Butene} \quad\quad\quad\quad\quad \text{2-Butene}$$

The names of substituents are then added as prefixes to the name of the alkene.

Compounds that contain C≡C triple bonds are called **alkynes**. These compounds have four less hydrogen atoms than the parent alkanes, so the generic formula for an alkyne with a single C≡C triple bond is C_nH_{2n-2}. The simplest alkyne has the formula C_2H_2 and is known by the common name acetylene.

$$H-C≡C-H \quad \text{Acetylene}$$

The IUPAC nomenclature for alkynes names these compounds as derivatives of the parent alkane, with the ending-yne replacing-ane.

$$CH_3C≡CCH_2CH_3 \quad \text{2-Pentyne}$$
$$HC≡CCH_2CH_3 \quad \text{1-Butyne}$$
$$\overset{CH_3}{\underset{|}{}}$$
$$CH_3C≡CCH_2CHCH_2CH_3 \quad \text{5-Methyl-2-heptyne}$$

In addition to compounds that contain one double bond (alkenes) or one triple bond (alkynes), we can also envision compounds with two double bonds (dienes), three double bonds (trienes), or a combination of double and triple bonds.

$$CH_3CH=CHCH_2C≡CH \quad \text{4-Hexen-1-yne}$$
$$CH_2=CHCH=CH_2 \quad \text{1,3-Butadiene}$$

Selected from "English for Chemistry and Chemical Engineering, by Zhang Yuping etc., Chemical Industry Press, 2007,156-164"

Lesson 23 Structure and Nomenclature of Hydrocarbons

New Worlds

1. nomenclature [nəˈmenklətʃər] n. 命名法，术语
2. saturated [ˈsætʃəreitid] a. 饱和的
3. alkane [ˈælkein] n. 链烷，烷烃
4. valence [ˈveiləns] n. （化合）价，原子价
5. tetravalent [ˌtetrəˈveilənt] a. 四价的; n. 四价染色体
6. methane [ˈmeθein] n. 甲烷，沼气
7. ethane [ˈeθein] n. 乙烷
8. repulsion [riˈpʌlʃən] n. 推斥，排斥，严拒
9. tetrahedral [ˈtetrəˈhedrəl] a. 四面体的
10. generic [dʒiˈnerik] a. 属的，类的，一般的，普通的
11. tar [tɑː] n. 焦油
12. skeleton [ˈskelitn] n. 骨架，骨骼，基干
13. branched a. 有枝的，分岔的
14. reminiscent [remiˈnis(ə)nt] a. 回忆往事
15. cyclopentane [ˌsaikləˈpentein] n. 环戊烷
16. planar [ˈpleinə] a. 平面的，平坦的
17. titled a. 倾斜的，翘起的
18. systematic [ˌsistiˈmætik] a. 体系的，系统的
19. derivative [diˈrivətiv] n. 衍生物
20. substituent [sʌbˈstitjuənt] n. 取代; a. 取代的
21. alphabetical [ˌælfəˈbetikəl] a. 依字母顺序的
22. alkene [ˈælkiːn] n. 烯烃，链烃
23. alkyne [ˈælkain] n. 炔
24. propane [ˈprəupein] n. 丙烷
25. butane [ˈbjuːtein] n. 丁烷
26. isobutene [ˌaisəuˈbjuːtein] n. 异丁烷
27. constitutional [ˌkɔnstiˈtjuːʃənəl] a. 构成的
28. pentane [ˈpentein] n. 戊烷
29. isopentane [aisəuˈpentein] n. 异戊烷, 2-甲基戊烷
30. neopentane [ˌniːəuˈpentein] n. 新戊烷
31. cycloalkane [ˌsaikləuˈælkein] n. 环烷
32. envision [inˈviʒən] vt. 想象，预想
33. cyclopropane [ˈsaikləuˈprəupein] n. 环丙烷
34. tetrahedron [ˈtetrəˈhedrən] n. 四面体
35. susceptible [səˈseptəbl] a. 易受影响的; n. （因缺乏免疫力而）易得病的人
36. cyclobutane [ˌsaikləuˈbjuːtein] n. 环丁烷
37. pucker [ˈpʌkə] v. 折叠 n. 皱纹
38. olefin [ˈəuləfin] n. 烯烃
39. crystallize [ˈkristəlaiz] vt.;vi. （使）结晶
40. hypothetical [ˌhaipəuˈθetikəl] a. 假设的，假定的，爱猜想的
41. unpaired [ʌnˈpɛəd] a. 不成双的，无对手的
42. suffix [ˈsʌfiks] n. 后缀，下标; vt. 添后缀
43. prefix [ˈpriːfiks] n. 前缀，（人名前的）称谓（如 Mr.Dr.Sir 等）
44. acetylene [əˈsetiliːn] n. 乙炔，电石气
45. diene [daiiːn] n. 二烯（烃）
46. triene [ˈtraiiːn] n. 三烯

Phrases

boiling point 沸点
tilt up 翘起
be thought of⋯ 被认为是⋯⋯
by the time 当⋯⋯时

Affixes

-ane ⋯⋯烷，如：propane 丙烷；methane 甲烷；ethane 乙烷
n-表示"正"，如：n-pentane 正戊烷
-ene ⋯⋯烯，如：enthene 乙烯；propylene 丙烯；butylene 丁烯
-yne ⋯⋯炔，如：ethyne 乙炔；alkyne 炔烃
-yl ⋯⋯，如：methyl 甲基；ethyl 乙烷基；cyclobutyl 环丁基
tri- 表示"三"，如：triene 三烯；triacetamide 三乙酰胺；triacid 三（酸）价的
tetra- 表示"四"，如：tetrahedral 四面体；tetrachloride 四氯化物；tetraborate 四硼酸盐

Notes

1. The generic formula for these compounds can be understood by assuming that they contain chains of CH₂ groups with an additional hydrogen atom capping either end of the chain. 系含有介词宾语从句（by assuming that⋯）的复合句，主句为被动语态。参考译文:我们可以把表中那些化合物的分子通式理解成这样，即它们包含由 CH₂ 基团组成的链，而在该链的两端都加上一个氢原子。

2. One of the four carbon atoms in the cyclobutane ring is therefore displaced from the plane of the other three to form a "puckered" structure that is vaguely reminiscent of the wings of a butterfly. 该复合句中定语从句 "that is vaguely⋯" 修饰不定式宾语 "structure"。参考译文：在环丁烷四个碳原子所组成的环中，其中一个碳并不在其他三个碳原子所确定的平面上，这样就形成了一种折起的结构，恰似我们想象中蝴蝶的两只翅膀。

3. One of these carbon atoms is titled up, out of the ringing, whereas the other is titled down to form the "chair" structure shown in Fig. 23-9, 文中 ring 指由其他四个碳原子所组成的圆环，one 和 the other 是指六元环中相对两端的两个碳原子。参考译文：（六元环中相对的两个碳原子中的）一个向上翘起，脱离（由其他四个碳原子所组成的）圆环平面，而相对的另一个碳原子向下翘，这样就形成了图 23-9 所示的 "椅式" 结构。

4. Because an alkene can be thought of as a derivative of an alkene from which an H₂ molecule has been removed, the genetic formula for an alkene with one C═C double bond is C_nC_{2n}. 参考译文：因为烯烃可以被认为是由烷烃派生出来，即由烷烃分子脱去一分子氢形成的，所以包含一个 C═C 双键的单烯烃的分子通式为 C_nC_{2n}。

Exercise

1. Put the following into Chinese:

    saturated   systematic   tilt up   crystallize
    prefix   alphabetical   substituent   unambiguous
    tetrahedron   alkane   envision   ethyl

2. Put the following into English:

    易受影响的   衍生物   沸点   烯烃
    后缀   结晶   假设的   命名法
    烷基   平面的   甲基   位次

3. Comprehension and toward interpretation

    a. For C_2H_5OH, alcohol is a common name or systematic name?
    b. Which names from the basis for organic compounds?
    c. Special attention should be paid to the prefixes and suffixes of organic compounds. Is that right?
    d. What are the uses of nomenclature rules?

Reading Material

Nomenclature of Chemical Compounds

The necessity of giving each compound a unique name requires a richer variety of terms that are available with descriptive prefixes such as n-and iso-. The naming of organic compounds is facilitated[1] through the use of formal systems of nomenclature. Nomenclature in organic chemistry is of two types: common and systematic. Common names for organic compounds originate in many different ways but share the feature that there is no necessary connection between name and structure. The name that corresponds to a specific structure must simply be memorized, much like learning the name of a person. Systematic names, on the other hand, are keyed directly to molecular structure according to a generally agreed upon sets of rules. The most widely used standards for organic nomenclature have evolved from suggestions made by a group of chemists assembled for that purpose in Geneva in 1892 and have been revised on a regular basis by the International Union of Pure and Applied Chemistry (IUPAC)[2]. The IUPAC rules govern all classes of organic compounds but are ultimately based on alkane names. Compounds in other families are viewed as derived from alkanes by appending functional groups to, or otherwise modifying, the carbon skeleton.

The IUPAC rules assign names to unbranched alkanes according to the number of their carbon atoms. Methane, ethane, and propane are used for CH_4, CH_3CH_3, and $CH_3CH_2CH_3$, respectively. The unbranched alkane $CH_3CH_2CH_2CH_3$ is defined as butane, not n-butane as given above. (The *n*-prefix is not used in systematic IUPAC nomenclature.) Beginning with five-carbon chains, the names of unbranched alkanes consist of a Latin or Greek stem corresponding to the number of carbons in the chain followed by the suffix –ane. Some examples are given in Table 23-2. A group of compounds, such as the unbranched alkanes, that differ from one another by successive introduction of CH_2 groups constitute what is called a homologous series.

Table 23-2 IUPAC Names of Unbranched Alkanes

Alkane formula	Name	Alkane formula	Name
CH_4	methane	$CH_3(CH_2)_6CH_3$	octane
CH_3CH_3	ethane	$CH_3(CH_2)_7CH_3$	nonane
$CH_3CH_2CH_3$	propane	$CH_3(CH_2)_8CH_3$	decane
$CH_3(CH_2)_2CH_3$	butane	$CH_3(CH_2)_{13}CH_3$	pentadecane
$CH_3(CH_2)_3CH_3$	pentane	$CH_3(CH_2)_{18}CH_3$	icosane
$CH_3(CH_2)_4CH_3$	hexane	$CH_3(CH_2)_{28}CH_3$	triacontane
$CH_3(CH_2)_5CH_3$	heptane	$CH_3(CH_2)_{98}CH_3$	hectane

Alkanes with branched chains are named on the basis of the name of the longest continuous chain of carbon atoms in the molecule, called the parent. The alkane shown has seven carbons in its longest continuous chain and is therefore named as a derivative of heptane, the unbranched alkane that contains seven carbon atoms. The position of the CH_3 (methyl) substituent on the seven-carbon chain is specified by a number (3-), called a locant, obtained by successively numbering the carbons in the parent chain starting at the end nearer the branch. The compound is therefore called 3-methylheptane.

$$\begin{array}{cccccccc} 1 & 2 & 3 & 4 & 5 & 6 & 7 \\ CH_3CH_2CHCH_2CH_2CH_2CH_3 & & & & & & & \text{3-Methylheptane} \\ & & | & & & & & \\ & & CH_3 & & & & & \end{array}$$

When there are two or more identical substituents, replicating prefixes (di-, tri-, tetra-, etc.) are used, along with a separate locant for each substituent. Different substituents, such as an ethyl ($-CH_2CH_3$) and a methyl ($-CH_3$) group, are cited in alphabetical order. Replicating prefixes are ignored when alphabetizing. In alkanes numbering begins at the end nearest the substituent that appears first on the chain so that the prefix numbers are as low as possible.

Methyl and ethyl are examples of alkyl groups. IUPAC nomenclature rules for naming alkyl groups beyond these two extend to cover even very complex structures. The IUPAC rules are unambiguous in the sense that there is no possibility that two different compounds will have the same name.

$$CH_3CHCH_2CCH_2CH_2CH_2CH_3 \quad \text{4-Ethyl-2,4-dimethyloctane}$$
(with CH₃, CH₃ substituents and CH₂CH₃ substituent)

Selected from "English for Chemical Engineers, by Ma Zhengfei etc., Southeast University Press, 2006, 150-152".

New Words

1. nomenclature ['nəumənkleiʃə] *n.* 命名法
2. alkane ['ælkein] *n.* 烷烃
3. prefix ['priːfiks] *n.* 前缀
4. suffix ['sʌfiks] *n.* 后缀
5. skeleton ['skelətn] *n.* 骨架，构架
6. stem [stem] *n.* 茎，柄
7. locant ['ləukənt] *n.* 位次
8. substituent [səb'stitjuənt] *n.* 取代基
9. alphabetical [ælfə'betikəl] *a.* 字母顺序的
10. alkyl ['ælkil] *n.* 烷基
11. ethyl ['eθil] *n.* 乙基
12. methyl ['meθil] *n.* 甲基
13. unambiguous [ʌnæm'bigjuəs] *a.* 明确的，清楚的

Notes

1. be facilitated 变得更为方便。
2. International Union of Pure and Applied Chemistry (IUPAC) 国际理论化学和应用化学联合会，其出版的刊物名称为"Pure and Applied Chemistry"。

Lesson 24
Green Science and Technology

Green Science

Although science is a widely used word having somewhat different meanings in different contexts, it can generally be regarded as a body of knowledge or system of study dealing with an organized body of facts verifiable by experimentation that are consistent with a number of general laws. In its purest sense, science avoids value judgments; it involves a constant quest for truths whether they be good, such as the biochemical basis of a cure for some debilitating disease, or bad, such as the nuclear physics behind the development of nuclear bombs. However, in defining green science, it is necessary to modify somewhat the view of "pure" science. *Green science* is science that is oriented strongly toward the maintenance of environmental quality, the reduction of hazards, the minimization of consumption of nonrenewable resources and overall sustainability.

When the public thinks of environmental pollution, exposure to hazardous substances, consumption of resources such as petroleum feedstocks, and other unpleasant aspects of modern industrialized societies, chemical science (the science of matter) often comes to mind[1]. So, it is fitting that to date the most fully developed green science is *green chemistry* defined as the practice of chemistry in a manner that maximizes its benefits while eliminating or at least greatly reducing its adverse impacts[2]. Green chemistry is based upon "twelve principles of green chemistry" and, since the mid-1990s, has been the subject of a number of books, journal articles, and symposia. In addition, centers and societies of green chemistry and a green chemistry journal have been established.

Green Technology

Technology refers to the ways in which humans do and make things with materials and energy directed toward practical ends. In the modern era, technology is to a large extent the product of engineering based on scientific principles. Science deals with the discovery, explanation, and development of theories pertaining to interrelated natural phenomena of energy, matter, time, and space. Based on the fundamental knowledge of science, engineering provides the plans and means to achieve specific practical objectives. Technology uses these plans to carry out the desired objectives. Technology obviously has enormous importance in determining how human activities affect Earth and its life support systems.

Technology has been very much involved in determining levels of human population on Earth, which has seen three great growth spurts since modern humans first appeared. The first of these, lasting until about 10,000 years ago, was enabled by the primitive, but remarkably effective tools that early humans developed, resulting in a global human population of perhaps 2 or 3 million. For example, the bow and arrow enabled early hunters to kill potentially dangerous game for food at some (safer) distance without having to get very close to an animal and stab it with a spear or club it into submission. Then, roughly 10,000 years ago, humans who had existed as hunter/gatherers learned to cultivate plants and raise do-masticated animals, an effort that was aided by the further development of tools for cultivation and food production. This development ensured a relatively dependable food supply in smaller areas. As a result, humans were able to gather food from relatively small agricultural fields rather than having to scout large expanses of forest or grasslands for game to kill or berries to gather. This development had the side effect of allowing humans to remain in one place in settlements and gave them more free time in which humans freed from the necessity of having to constantly seek food from their natural surroundings could apply their ingenuity in areas such as developing more sophisticated tools. The agricultural revolution allowed a second large increase in numbers of humans and enabled a human population of around 100 million, 1000 years ago. Then came the industrial revolution, the most prominent characteristic of which was the ability to harness energy other than that provided by human labor and animal power[3]. Wind and water power enabled mills and factories to use energy in production of goods. After about 1800, this power potential was multiplied many fold with the steam engine and later the internal combustion engine, turbines, nuclear energy, and electricity, enabling current world population of around 6 billion to grow (though not as fast as some of the more pessimistic projections from past years).

There is ample evidence that new technologies can give rise to unforeseen problems. According to the law of unintended consequences, whereas new technologies can often yield predicted benefits, they can also cause substantial unforeseen problems. For example, in the early 1900s, visionaries accurately predicted the individual freedom of movement and huge economic boost to be expected from the infant automobile industry. It is less likely that they would have predicted millions of deaths from automobile accidents, unhealthy polluted air in urban areas, urban sprawl, and depletion of petroleum resources that occurred in the following century. The tremendous educational effects of personal computers were visualized when the first such devices came on the market. Less predictable were the mind-numbing hours that students would waste playing senseless computer games. Such unintended negative consequences have been called revenge effects. Such effects occur because of the unforeseen ways in which new technologies interact with people.

Avoiding revenge effects is a major goal of green technology defined as technology applied in a manner that minimizes environmental impact and resource consumption and maximizes economic output relative to materials and energy input[4]. During the development phase, people who develop green technologies, now greatly aided by sophisticated computer methodologies, attempt to predict undesirable consequences of new technologies and put in place preventative measures before revenge effects have a chance to develop and cause major problems.

A key component of green technology is industrial ecology, which integrates the principles of science, engineering, and ecology in industrial systems through which goods and services are provided, in a way that minimizes environmental impact and optimizes utilization resources, energy, and capital. In so doing, industrial ecology considers every aspect of the provision of goods and services from concept, through production, and to the final fate of products remaining after they have been used. It is above all a sustainable means of providing goods and services[5]. It is most successful in its application when it mimics natural ecosystems, which are inherently sustainable by nature. Industrial ecology works through groups of industrial concerns, distributors, and other enterprises functioning to mutual advantage, using each others' products, recycling each others' potential waste materials, and utilizing energy as efficiently as possible. By analogy with natural ecosystems, such a system comprises an industrial ecosystem.

Selected from "Stanley E Manahan, Environmental Science and Technology, A Sustainable Approach to Green Science and Technology, Taylor & Franicis Group, USA, 2007"

New Words

1. verifiable ['verifaiəbl]   *n.* 可核实的
2. symposia [sim'pəuziə]   *n.* 专题报告会，讨论会，专题文集
3. pertaining [pə'teiniŋ]   *a.* 附属……的，与……有关的，为……所固有的
4. spurt [spə:t]   *vt.; vi.* 喷射出，急剧上升，生长，发芽
5. stab [stæb]   *vt.; vi.* 刺(穿，伤)，企图，努力，尝试
6. scout [skaut]   *vt.; vi.* 侦察，搜索，监视，发现，排斥
7. harness ['ha:nis]   *vt.* 控制，治理，利用，开发
8. unintended consequence   缺乏研究（知识，经验，外行）的结果(结论)
9. visionary ['viʒə‚neri:]   *a.; n.* 幻想的，空想，非实际
10. mind-numbing   失去感觉的心情（头脑），迟钝的感觉神志（头脑）
11. revenge effect   报复效应
12. mimic ['mimk]   *a.; vt.; n.* 模仿的，假装的，仿造物（品）

Notes

1. When the public thinks of environmental pollution, exposure to hazardous substances, consumption of resources such as petroleum feedstocks, and other unpleasant aspects of modern industrialized societies, chemical science (the science of matter) often comes to mind. 参考译文：当公众想起环境污染时，如有害物质的暴露、石油原料的消耗、现代工业社会其他令人不愉快的方面，化学学科（物质科学），常会映入脑中。

2. So, it is fitting that to date the most fully developed green science is green chemistry defined as the practice of chemistry in a manner that maximizes its benefits while eliminating or at least greatly reducing its adverse impacts. 参考译文：因此，迄今为止，最成熟的绿色科学是绿色化学，它定义为以最大限度排除或削减其不利影响时获得最大利益方式的化学实践。

3. Then came the industrial revolution, the most prominent characteristic of which was the ability to harness energy other than that provided by human labor and animal power. 参考译文：接着，工业革命来临，工业革命重要的特征是能够利用能量，而不是由人力或动物提供能量（该句为倒装句，因为主语太长，把动词 came 放在主语前面）。

4. Avoiding revenge effects is a major goal of green technology defined as technology applied in a manner that minimizes environmental impact and resource consumption and maximizes economic output relative to materials and energy input. 参考译文：避免环境报复效应是绿色技术的主要目的，绿色技术定义为使得环境影响和资源消耗最小，相对于物质与能量输入而言，经济输出最大的技术。

5. It is above all a sustainable means of providing goods and services. 参考译文：最重要的是提供物质与服务的可持续手段。

Exercise

1. Put the following into Chinese：

harness   automobile accidents   pertaining   prominent

symposia	mind-numbing	feedstocks	green chemistry
verifiable	unintended consequence	consumption	hunter/gatherers

2. Put the following into English:

专题报告会	缺乏经验的结果	原料	报复效应
工业生态学	无关的自然现象	幻想	利用能量
资源消耗	预防措施	模仿的	环境影响

3. Comprehension and towards interpretation

　　a. What are the definition of Science and Technology based on the text?
　　b. What are Green Technology and Industrial Ecology according to the text?
　　c. What is the definition of Green Chemistry?

Reading Material

Green Chemistry

　　Green chemistry, also called sustainable chemistry, is a philosophy of chemical research and engineering that encourages the design of products and processes that minimize the use and generation of hazardous substances. Whereas environmental chemistry is the chemistry of the natural environment, and of pollutant chemicals in nature, green chemistry seeks to reduce and prevent pollution at its source.

　　As a chemical philosophy, green chemistry applies to organic chemistry, inorganic chemistry, biochemistry, analytical chemistry, and even physical chemistry. While green chemistry seems to focus on industrial applications, it does apply to any chemistry choice. Click chemistry is often cited as a style of chemical synthesis that is consistent with the goal of green chemistry. The focus is on minimizing the hazard and maximizing the efficiency of any chemical choice.

　　Green chemistry protects the environment, not by cleaning up, but by inventing new chemist and new chemical processes that do not pollute. Green chemistry emphasizes renewable starting materials for a bio-based economy.

　　（1）Twelve Principles of Green Chemistry

　　① Prevention

　　It is better to prevent waste than to treat waste after it has been created.

　　② Atom Economy

　　Synthetic methods should be designed to maximize the incorporation of all materials used in the process into the final product.

　　③ Less Hazardous Chemical Syntheses

　　Wherever practicable, synthetic methods should be designed to use and generate substances that possess little or no toxicity to human health and the environment.

　　④ Designing Safer Chemicals

　　Chemical products should be designed to effect their desired function while minimizing their toxicity.

　　⑤ Safer Solvents and Auxiliaries

　　The use of auxiliary substances (e. g., solvents, separation agents, etc.) should be made unnecessary wherever possible and innocuous when used.

⑥ Design for Energy Efficiency

Energy requirements of chemical processes should be recognized for their environmental and economic impacts and should be minimized. If possible, synthetic methods should be conducted at ambient temperature and pressure.

⑦ Use of Renewable Feedstocks

A raw material or feedstock should be renewable rather than depleting whenever technically and economically practicable.

⑧ Reduce Derivatives

Unnecessary derivatization (use of blocking groups, protection/deprotection, temporary modification of physical/chemical processes) should be minimized or avoided if possible, because requiring additional reagents and can generate waste.

⑨ Catalysis

Catalytic reagents (as selective as possible) are superior to chemical reagents.

⑩ Design for Degradation

Chemical products should be designed at the end of their function they break down into innocuous degradation products and do not persist in the environment.

⑪ Real-time Analysis for Pollution Prevention

Analytical methodologies need to be further developed to allow for real time, in-process monitoring and control prior to the formation of hazardous substances.

⑫ Inherently Safer Chemistry for Accident Prevention

Substances and the form of a substance used in a chemical process should be chosen to minimize the potential for chemical accidents, including releases, explosions, and fires.

(2) **Green Chemical Products**

The application of green chemistry concepts is becoming more widespread. The Green Chemistry and Consumer Network in the UK, for instance, alerts retailers and consumers around the world to new developments in safer product design.

① Green Paints

Many now recognize that volatile organic compounds (VOCs), the sources of paint smell, are harmful to health and the environment. Great strides have been made to bring home paints to the market that contain low or no VOCs. One company, Archer RC Paint, won a 2005 Presidential Green Chemistry Award with a bio-based paint which in addition to lower odor, has better scrub resistance and better opacity.

② Green Plastics

Some plastic products can be made from plant sugars from renewable crops, like corn, potatoes and sugar beets instead of non-renewable petroleum. For example, the U. S.-based company Nature Works LLC markets a bio-based polymer PLA, from corn which is used in food and beverage packaging, as well as a 100% corn fiber, ingeo, which is used in blankets and other textiles. Interface Fabrics uses PLA in their fabrics but also carefully integrates green chemistry principles when choosing dyes for their PLA based product lines[1]. A collaboration of group have produced Sustainable Bio-materials Guidelines that outline a comprehensive sustainable life cycle approach from agricultural practices through to end of life recycling and composting[2].

③ Green Carpets in All Sorts of Places

In 2003, Shaw Carpet won a Presidential Green Chemistry Challenge Award with its carpet tile backing, EcoWorx replaces conventional carpet tile backings that contain bitumen, polyvinyl chloride (PVC) or polyurethane with ployolefin resins which have low toxicity. This product also provides better adhesion, does not shrink, and can be recycled. Carpets with EcoWorx backing are now available for our homes, schools, hospitals and offices.

These are just a few examples which demonstrate how some companies are integrating green chemistry principles into their product design.

（3）**Trends of Green Chemistry**

In 2005, American Nobel Prize winner R. Noyori identified three key developments in green chemistry the use of super-critical carbon dioxide as green solvent, hydrogen peroxide for clean oxidation and the use of hydrogen in asymmetric synthesis. Examples of applied green chemistry are super-critical water oxidation, on water reactions, and dry media reactions.

Bioengineering is also seen as a promising technique for achieving green chemistry goals. A number of important process chemicals can be synthesized in engineered organisms, such as shikimate, a Tamiflu precursor which is fermented by Roche in bacteria[3].

Attempts are being made not only to quantify the greenness of a chemical process but also to factor in other variables such as chemical yield, the price of reaction components, safety in handling chemicals, hardware demands, energy profile and ease of product workup and purification[4].

In one quantitative study, the reduction of nitrobenzene to aniline receives 64 points out of 100 marking it as an acceptable synthesis overall whereas a synthesis of an amide using HMDS is only described as adequate with a combined 32 points.

Green chemistry is seen as a powerful tool that researchers must use to evaluate the environmental impact of nanotechnology. As nanomaterials and the processes to make them must be considered to ensure their long-term economic viability[5].

Selected from "English for Chemical Engineering and Technology, by Jia Changying etc., China Petrochemical Press, 2018, 136-139"

New words

1. aniline [ˈænəˌlin]   n. 苯胺
2. bitumen [bəˈtuːmən]   n. 沥青，柏油
3. composting [ˈkaːmpəustiŋ]   n. 堆制肥料
4. ingeo [inˈdʒiəu]   n. 聚乳酸纤维
5. opacity [uəˈpæsəti]   n. 不透明性
6. polyurethane [ˌpɔliˈjuərəθein]   n. 聚氨基甲酸酯
7. polyvinyl [pɔlivainl]   n. 乙烯基聚合物
8. shikimate [ˈʃikmeit]   n. 莽草酸盐
9. carpet tile backing   n. 地毯背衬
10. HMDS (Hexamethyl Disilazane)   n. 六甲基二硅氮烷

Notes

1. Interface Fabrics uses PLA in their fabrics but also carefully integrates green chemistry principles when choosing dyes for their PLA based product lines. 参考译文： Interface Fabrics 公司在他们的纺织物中使用聚乳酸，并在为基于聚乳酸的产品线选择染料时注意结合绿色化学原则。

2. A collaboration of group have produced Sustainable Bio-materials Guidelines that outline a comprehensive sustainable life cycle approach from agricultural practices through to end of life recycling and composting. 参考译文：公司合作共同制定了可持续发展材料指南，该指南勾勒出一个全面的可持续发展的生物循环方法，该方法贯穿农业耕作到生命周期的结束和堆肥。

3. Bioengineering is also seen as a promising technique for achieving green chemistry goals.A

number of important process chemicals can be synthesized in engineered organisms, such as shikimate, a Tamiflu precursor which is fermented by Roche in bacteria. 参考译文：生物工程技术可实现绿色化学目标。许多重要化学品都可以在工程化的有机体中合成，如莽草酸盐就是Roche公司在细菌中发酵生成的达菲前体。

4. Attempts are being made not only to quantify the greenness of a chemical process but also to factor in other variables such as chemical yield, the price of reaction components, safety in handling chemicals, hardware demands, energy profile and ease of product workup and purification. 参考译文：人们正尝试不仅要量化化学工程的绿色环保，而且要将绿色环保技术分解成其他变量，如化学产率、反应组分的价格、化学品的使用安全性、硬件需求、能源概况以及产品检查和纯化的简易性。

5. Green chemistry is seen as a powerful tool that researchers must use to evaluate the environmental impact of nanotechnology. As nanomaterials and the processes to make them must be considered to ensure their long-term economic viability. 参考译文：绿色化学被视为研究人员评估纳米技术对环境影响时必须使用的一种强大工具。随着纳米材料的开发，纳米产品和生产过程对环境和人类健康的影响必须纳入考虑范围，以确保产品长期的经济可行性。

Lesson 25
Processing and Utilization of Coal

Coal

Coal is the organic mineral precipitate transformed from the chemical and physical reaction of the plant remains, which is the complex mixture of organic and inorganic substances. The main component of coal is organic substance. There are great differences among different kinds of coal, different parts of the same kind of coal, different coal seams of the same region. The reason for the various differences are closely related to the substances, environment and reactions of the materials that form the coal.

Plants are mainly composed of organic substances, but also contain a certain amount of inorganic substances. From the chemical point of view, the organic substances can be divided into four groups: sugar and its derivatives, lignin, proteins and lipids, but not all the plants can be transformed into coal, the formation of the coal has four conditions, namely: the ancient plant species; climatic conditions; natural and geographical conditions and the conditions of crystals movement and only with these four conditions at the same time, large reserves of the major coalfields can be transformed after their long and harmonious reaction. The formation of coal is a complex and lengthy process; it normally takes millions of years to hundreds of millions of years. Plants are gradually transformed into coal, from junior stage to senior stage, which are: plants, peat, lignite, bituminous coal, anthracite.

Coal, oil, and natural gas are the primary fossil energy. According to the forecasts of China National Petroleum Corporation, from 2010 to 2020, fossil energy will remain the dominant position in the global primary energy consumption. Based on the characteristics of coal resource, it is obvious that coal resource will remain China major energy consumption in the decades.

Natural Gas

Natural Gas is a mixture of multi-component gas. Its main component is hydrocarbons, among which methane is the predominant one in addition to small amounts of ethane, propane and butane. Moreover, there are also hydrogen sulfide, carbon dioxide, nitrogen, and water vapor and traces of inert gas such as helium and argon. Under standard condition, methane to butane exist in the state of gas, above pentane are liquid. The matters which can affect the health of respiratory that natural gas produced in the process of combustion are quite rare. The amount of carbon dioxide natural gas produces is only about 40% of that of coal produces, releasing very little sulfur dioxide. What's more, after combustion of natural gas, there are no residue and waste water produced. Compared with coal, oil and other energy resources, natural gas possess the advantages of the use of safe, high heat valve, cleanliness and other advantages.

Processing and Utilization of Coal

Coal is an important energy source, also an important raw material in metallurgy and chemical industry. which is mainly used in combustion, carbonization, gasification, liquefaction and soon.

(1) **Combustion**

The purpose of burning coal should be to release their energy maximumly and quickly. In the cumbustion process of coal, it generates gas compound and the residues of solid carbon. The simplest gas compounds are methane and acetylene, and water gas can be produced at high temperature. The coal combustion process is chemical reaction that happens on the surface of coal. The reaction speed depends on the size of the coal surface and the quantity of supplied air, the larger the surface area of coal is, the more the amount of supplied air, then the faster the speed will respond.

(2) **Coal Carbonization**

It refers to the process in which coal is heated and decomposed in the absence of air, then generates coke (or semi-coke), coal tar, crude benzene, gas and other products. According to different final temperature at different heating condition, coal carbonization can be divided into three types: 900~1100℃ high temperature carbonization, also namely the coking; 700~900℃ medium temperature carbonization, 500~600℃ low temperature carbonization. High temperature carbonization (coking) is widely used. Its product, coke, is mainly used for blast furnace iron-making and foundry, and also can be used to produce nitrogen fertilizer and calcium carbide; coke oven gas is a kind of fuels, also an important chemical raw material. Coal tar can be used to produce fertilizer, pesticides, svnthetic fiber, svnthetic rubber, paints, dyes, pharmaceuticals, explosives and so on.

(3) **Coal Gasification**

It is a thermo-chemical process in which coal or coal coke as raw material, oxygen (air, enrichment of oxygen or pure oxygen), steam or hydrogen as gasification agent, the combustible part of coal coke is transformed into the fuel gas at high temperature by chemical reaction. The coal gasification process can be divided into five basic types: self-heated steam gasification, outside heated steam gasification, hydrogenation of coal gasification, the substitute of natural gas by manufacturing the combination of coal steam gasification and hydrogenation gasification, the substitute of natural gas by manufacturing the combination of coal steam gasification and methanation.

(4) **Coal Liquefaction**

It is an advanced clean coal technology by which the solid coal is turned into liquid fuels, chemical raw materials and products by chemical process. Coal liquefaction can be divided into direct liquefaction and indirect liquefaction. Direct liquefaction is to pyrolysis and hydrogenate coal molecules in the role of catalyst and solvent at high temperature (400℃ above) and high pressure (10MPa or above), directly converting into liquid fuel, and then further processed and refined into gasoline, diesel and other fuel oil. Coal liquefaction is known as hydroliquefaction. Indirect liquefaction of coal is a process in which all the coal is firstly gasified into synthetic gas, and then takes the coal-based synthetic gas (carbon monoxide and hydrogen) as raw materials which are catalyzed and synthesized into hydrocarbon fuel, chemical raw materials and products.

Selected from "English for Chemical Engineering and Technology, by Jia Changying etc., China Petrochemical Press, 2018, 106-109"

New words

1. argon ['ɑːgɔn]   *n.* 氩
2. anthracite ['ænθrəsait]   *n.* 无烟煤；硬煤
3. bituminous [biˈtjuːminəs]   *adj.* 含沥青的
4. fossil ['fɔsl]   *n.* 化石

5. lengthy ['leŋθi]   adj. 漫长的
6. lipids ['lipidz]   n. 脂类
7. lignin ['lignin]   n. 木质素
8. lignite ['lignait]   n. 褐煤
9. mineral ['minərəl]   n. 矿物质
10. peat [pi:t]   n. 泥煤，泥炭
11. precipitate [pri'sipiteit]   n. 沉淀物
12. respiratory [rə'spɪrətri]   adj. 呼吸的
13. seam [si:m]   n. 煤层
14. high heat valve   热值高
15. blast furnace   n. 鼓风炉
16. combustible [kəm'bʌstəbl]   adj.; n. 易燃的; 易燃物
17. crude benzene   n. 粗苯
18. foundry ['faundri]   n. 铸造
19. liquefaction [,likwi'fækʃən]   n. 液化
20. metallurgy [mə'tælədʒi]   n. 冶金(学)
21. pesticide ['pestisaid]   n. 杀虫剂，农药
22. calcium carbide   电石
23. coal tar   煤焦油；煤黑油
24. coke oven gas   焦炉煤气
25. blast furnace iron-making and foundry   高温炼铁和铸造
26. self-heated steam gasification   自热式水蒸气气化
27. outside heated steam gasification   外热式水蒸气气化
28. combination of coal steam gasification and methanation   煤的水蒸气气化与甲烷化结合

Exercise

1. Put the following into Chinese：

   pesticide   hydroliquefaction   mineral   precipitate
   combustible   methanation   lengthy   crude benzene
   lipids   respiratory   lignin   coal tar

2. Put the following into English：

   冶金学   焦炉煤气   电石   高温炼铁
   热值高   铸造   惰性的   精炼
   无烟煤   加氢液化   合成纤维   气化

Reading Material

Summary of Coal Chemical Technology

Coal is the most abundant fossil resources in the world. In view of the situation of the rising oil price and promoting environmental protection, to develop the coal chemical industry, especially the new coal chemical industry and adjust our energy and chemical structure become increasingly important.

（1）**Coal Chemical Industry**

Coal chemical industry takes coal as raw material; change coal into gas, liquid, solid fuels and chemicals, and produce a variety of chemical products through chemical process Coal chemical industry includes primary chemical processing of coal, secondary chemical processing of coal and deeply chemical processing, the coking, gasification, liquefaction of the coal, synthetic gas process, tar chemical process and calcium carbide acetylene chemical process of coal and so on. According to the production process and different products, coal chemical industry can be mainly divided into coking, calcium carbide, coal gasification, coal liquefaction four main production chains, among which ammonia in coal coking, calcium carbide and coal liquefaction belong to the traditional coal chemical industry, while alcohol ether fuel cranked out through coal gasification and olefine cranked out through coal liquefaction, coal gasification and so on belong to the modern coal chemical industry.

(2) **Coal Chemical Technology**

① Coking

The process in which coal is isolated from the air and decomposed under heat-flash is also know as coal carbonization. The main products of coal carbonization are coal coking, coal tar (benzene, toluene, etc.), coke oven gas (hydrogen, methane, ethylene, carbon monoxide, etc.), and refined ammonia. All these products have been widely used in chemical, medicine, dyes, pesticides and carbon industry.

② Coal Gasification

Coal gasification is the thermal process in which solid carbon is converted into combustible gas (gas mixture) with chemical agents under high temperature. In this process, air, water vapor and carbon dioxide are taken as he gasification agent. The reaction they occur with the carbon in the coal is known as heterogeneous reaction.

③ Coal Liquefaction

The so-called coal liquefaction is to convert organic matter of coal into fluidity products, whose purpose is to get and use liquid hydrocarbons to replace petroleum and its products, including the two technologies of direct liquefaction and indirect liquefaction. Coal liquefaction products have huge market potential and ask for high concentration in terms of process and engineering, which has become an important developing direction of China's new coal chemical technology and industry.

Ⅰ Direct liquefaction

The direct liquefaction of coal was firstly invented by German scientist F. Bergius in 1913, whose principle is that the coal directly reacts with gaseous hydrogen with solvent under high temperature and pressure to increase the content of hydrogen in the coal and then turn into liquid finally.

Ⅱ Indirect Liquefaction

The indirect liquefaction of coal was firstly proposed by F. Ficher and H. Tropsch, the two chemists of German Royal Institute of Coal in 1923, so it is also known as F. Ficher-H. Tropsch (FT for short). The principle of the indirect coal liquefaction is that firstly synthesize gas ($CO+H_2$) based on coal, then based on the gas synthesize liquid hydrocarbon products in the function of catalyst.

Selected from "English for Chemical Engineering and Technology, by Jia Changying etc., China Petrochemical Press, 2018, 109-110"

Lesson 26
Reading and Searching a Patent

Patent is an official license from the government giving one person or business the exclusive right to make, use, and sell an invention for a limited period. Ideas are not eligible, neither is anything not new. The earliest known patent for an invention in England is dated 1449.

Reviewing patent documents requires the skill of understanding the significance of what is being disclosed. Legal counsel should always assist in interpreting the legal effect of any patent on commercial activity. However, a patent attorney or agent often must seek the assistance of technical personnel to gain a full understanding of the technology disclosed and claimed in a given patent. Further, an understanding of the form, content, and function of the various sections of a U.S. patent assists the nonlawyer in understanding the commercial importance of any issued patent.

The cover or front page of an U.S. patent must follow the form requirements placed on issued patents by the U.S.PTO. Specifically, the front cover discloses the inventor in two locations. The first named inventor is generally used as a head note for the patent. A given patent may often be referred to in an informal sense by this inventor's name.

Once the patent is issued, the inventor is referred to as the *patentee*. The first named inventor, if there is more than one, is printed prominently in the upper left-hand corner of the front page of the patent. All of the inventors or patentees are listed beneath the *invention title* along with the inventor's full names, addresses, and citizenship if other than the United States.

The title of the invention is generally written so as to use the shortest possible accurate description of the invention described fully in the patent and found in the claims. The *patent application number* and *filing date* are printed beneath the title. The application number and filing date are important because the patent application filing date may be used to eliminate other publications of third parties that might be used to limit the legal scope of the applicant's rights.

Also printed on the front page of the patent is *a coded classification listing*. This coding is complex and largely unnecessary to a lay person's understanding of a patent. This classification stems from the specific technology area to which the patent application was assigned during processing in the U.S.PTO. The classification also results from the search or review of prior patents completed by the Patent Examiner.

Apart from the technical classification information, the front page of the patent also contains a listing of publications or references cited during examination, including "United States Patent Documents", "Foreign Patent Documents", and "Other Publications" such as trade literature, journal articles, and product descriptions.

The front cover of the patent generally also identifies the U.S. Patent Examiner who reviewed and allowed the patent application, as well as the patent attorney, agent, or firm who worked with the Patent Examiner on the application.

Also provided is an *abstract*, which describes the invention, specifically highlighting its most valuable properties and distinguishing features. By doing so, the abstract assists those searching for prior patents which disclose developments relevant to an invention or patent application presently under examination. Another aid to patent searchers is the listing of claims and drawing sheets. *A representative drawing* may also often be found on the front page of the patent, if figures are provided by the inventor. Figures or drawings are not required to receive a patent. However, where figures are essential to a full and complete understanding of the invention, they must be included. Further, the figures should show those elements of the invention which are found in the claims.

Within the body of the issued patent, the title is generally repeated to maintain clarity. A field of invention is then provided. The field of invention should direct the reader to the general area of technology to which the invention relates, and to specific improvements in the identified areas of application.

A description or explanation of the background of the invention may also be provided by the inventor. This background section discusses previous developments of inventors working in the same area of technology, and may also list publications or patents that have discussed these developments and predate the filling date of the patent application. The background section may also point to deficiencies in the prior developments that the inventor intends to overcome.

To complement the discussion of problems and prior publication in the background of the invention, the inventor may generally provide a summary of the invention disclosed in insant patent. The summary of the invention should provide an explanation of the invention in the broadest and simplest terms and should also discuss how the invention disclosed in the patent solves problems remaining in prior work in this area of technology.

The patent should also provide a brief description of any drawings or figures. This brief description is often given in the technical terms used by engineering draftsmen to explain the various views illustrated in the figures.

The next section of the patent is titled "The Detailed Description of the Preferred Embodiment", often a multipage work serving several functions. First, the detailed description should provide an illustration of the invention in both its broadest or simplest sense and in its most preferred sense. Any elements of the invention that the inventor believes are crucial to the success or performance of the invention must also be included within this description. Further, this description should provide an explanation of the invention that is definite and illustrative, so as to allow persons having nothing but the patent before them to practice or use the invention in the manner intended. This description should be understood by those who work in the area that covers the subject matter of the patent.

Elements often include a detailed explanation of the various elements of the invention comprising the function of those elements, a written description of those elements, and an analysis of the elements that relies on any figures present in the patent application. The Detailed Description of the Preferred Embodiment may also include one or more working examples, especially if the invention is related to chemical technology. That is, in cases relating to chemistry, biochemistry, and chemical engineering, working examples are more often included than not. These working examples may serve any number of functions, including illustrating the formulation, applicability, and performance of the invention. Working examples may also be used to illustrate how the invention is distinguishable from those inventions previously developed and patented. As such, these working examples may include data such as adhesion and cohesion performance for adhesives, disinfecting and sanitizing efficacy for cleaners, or data on chemical and physical properties for polymer systems.

The final section of an issued patent is the claims. A United States patent is required by law to have at least one claim. The claims represent the legal definition and boundaries of the rights resulting from the patent grant. Patent claims are analogous to the legal description which one might find on a title to real estate.

When evaluating an issued patent for purposes of determining the patentability of a new invention, the entire patent must be considered. As a result, the figures and The Detailed Description of the Preferred Embodiment are every bit as important to an issued patent as the claims. At certain times any one of these elements may become more relevant than another. For example, claims tend to be more relevant to determinations of patent infringement or violation. However, in determinations concerning the patentability of new inventions, the figures and The Detailed Description of the Preferred Embodiment may be the most relevant aspects of any previous patent.

Patent Searches. Because valid patent claims can only be issued on an invention that is novel and innovative in light of prior art, it is necessary to search the prior art for previous references either to the composition of matter, press, or machine defined in the claims of a patent application, or to any similar composition, process, or apparatus that would render the claimed invention obvious to a person skilled in the field of the invention. Inventions that have been described in a publication or embodied in a product are said to have been anticipated in the prior art and are not patentable. Patentability searches are performed by examiners employed by the national and regional patent offices and are an important step in the examination of patent applications. Patentability searches should also be performed by the representatives of inventors prior to the filing of a patent application so that the claims will not overlap with any publication in the prior art. These searches may encompass the full scope of the published literature, including patents, technical journals, gray literature, and even catalogs. Individuals or organizations who are making plans to introduce a new product or process must conduct infringement searches to ensure that they will not infringe patents that belong to others. Infringement searches need only consider patents in force and pending applications that may result in patents in countries where manufacturing or marketing are contemplated. After a patent application has been published and/or a patent has been granted, organizations that wish to practice the invention may also conduct validity searches to be used as ammunition for opposition proceedings or invalidity lawsuits. Validity searches, like patentability searches, should include all forms of published literature, but are limited to publications with effective dates earlier than the filing date of the patent application being challenged.

Searches of scientific and technical literature are performed using any of the information retrieval tools suitable for searches done for other purposes. Patent offices have devised special classification systems to facilitate searches among the individual patent documents in their collections. These patent classification systems were designed to subdivide patents into groups covering similar inventions were claimed in later applications. All of the existing fields of science and technology were defined and provided with a class code and subdivisions of the fields were given narrower classification designations. Patents belonging to each subclass were stacked together in drawers or on shelves similar to the stacks of boxes in a shoestore, and examiners or members of the public could extract a stack of patents and search for information in the subfield of interest by flipping through paper copies of the patent documents. As new field of science and technology have developed, each patent classification system has been revised so that the emerging technologies can be searched. Patents are assigned classification codes by the examining office and the relevant primary classification and any cross-reference classifications are printed on the first page of the patent. Although patent classifications originated as tools for manual searches, they can be searched through printed or electronic indexes as well.

Patent systems were conceived encouraging the dissemination of information on technological developments. Information dissemination is therefore essential for the patenting process. Patent offices have traditionally announced the issuance of new patents in bulletins and gazettes. Other organizations, notably scientific and technical societies and for profit publishers, have produced value-added patent information services. These secondary sources of patent information serve multiple purposes, among which are current awareness alerting, document delivery, and retrospective searching. Traditionally, such products have appeared as printed publications, but increasingly they have found second use in electronic form in on-line databases, and in the 1990s there has been rapid growth of optical storage of information, especially as Compact Disk-Read Only Memory (CD-ROM) products. Patent documentation is a field in considerable ferment, with rapid introduction of new products and capabilities.

Printed Patent Office Gazettes. The issuance of patents is announced by patent offices in publications typically known as gazettes and bulletins, which are published most commonly at the time of the patent's publication, but there are exceptions. Advance information is published in a patent gazette by some countries prior to the publication of patent documents, typically as a notification of filing details. However, some patent gazettes do not appear until well after the effective publication date of the patents they announce. The amount of information included in patent gazettes varies. Typically, they include bibliographic details on published patent applications and granted patents, including patent number, title, inventor, patentee, patent classification, application number and date, and priority application details if relevant. Some gazettes also provide the front page abstract of the patent and a representative drawing. In addition to announcement of new patents and applications, the various gazettes typically include listing of patents that have been rejected, challenged, or disclaimed, patents that have been allowed to lapse, and in some instances even listings of new applications that have been made but that will not be published for some time, if ever. Gazettes often include indexes to the information they contain; the amount of indexing available varies from country to country.

Information from Other Sources. Some of the abstracting and indexing services produced by scientific and technical societies have traditionally included patent information, especially in the field of chemistry. For instance, *Chemical Abstracts* (CA), produced by the American Chemical Society since 1907, has always covered patents. On the other hand, some notable information services have not included patent coverage. One example, despite the fact that many patents are based on some aspects of engineering, is the *Engineering Index. Science Abstract*, covering physics, electricity, and electronics, is another example, which has not covered patents since 1976. However, even where patents are covered, the focus may not be ideal for those concerned with the legal aspects of patents. Thus, CA in its patent coverage documents the new chemistry involved, but shies away from the legal aspect of patents. For these and other reasons, others have stepped in to develop a variety of patent information services, e.g., Derwent information Ltd. of London.

Selected from "Encyclopedia of Chemical Technology, Vol. 19, R. E. Kirk and D. F. Othmer,Interscience, 3rd edition 1996"

New Words

1. attorney [əˈtɜːni] *n.* 代理人，律师
2. U.S. PTO=United States Patent Office 美国专利局
3. patentee [ˌpætənˈtiː] *n.* 专利权所有人
4. lay [leɪ] *a.* 外行的，局外的
5. draftsman=draughtsman [ˈdrɑːftsmən] *n.* 制图员，绘图员，起草者
6. embodiment [ɪmˈbɒdɪmənt] *n.* 具体化，具体体现
7. infringement [ɪnˈfrɪndʒmənt] *n.* 侵害，违反
8. gazette [gəˈzet] *n.* 公报，报纸
9. lapse [læps] *n.;v.* 失效，终止

Reading Material

Design Information and Data

Information on manufacturing processes, equipment parameters, materials of construction, costs and the physical properties of process materials are needed at all stages of design; from the initial screening of possible processes, to the plant start-up and production.

When a project is largely a repeat of a previous project, the data and information required for the design will be available in the company's process files, if proper detailed records are kept. For a new project or process, the design data will have to be obtained from the literature, or by experiment (research laboratory and pilot plant), or purchased from other companies. The information on manufacturing processes available in the general literature can be of use in the initial stages of process design, for screening potential process; but is usually mainly descriptive, and too superficial to be of much use for detailed design and evaluation.

The literature on the physical properties of elements and compounds is extensive, and reliable values for common materials can usually be found. Where values cannot be found, the data required will have to be measured experimentally or estimated. Methods of estimating (predicting) the more important physical properties required for design are given in this chapter.

Readers who are unfamiliar with the sources of information, and the techniques used for searching the literature, should consult one of the many guides to the technical literature that have been published.

1. Sources of Information on Manufacturing Processes

In this section the sources of information available in the open literature on commercial processes for the production of chemicals and related products are reviewed. The chemical process industries are competitive, and the information that is published on commercial process is restricted. The articles on particular processes published in the technical literature and in textbooks invariably give only a superficial account of the chemistry and unit operations used. They lack the detailed information needed on reaction kinetics, process conditions, equipment parameters, and physical properties needed for process design. The information that can be found in the general literature is, however, useful in the early stages of a project, when searching for possible process routes. It is often sufficient for a flow-sheet of the process to be drawn up and a rough estimate of the capital and production costs made.

The most comprehensive collection of information on manufacturing processes is probably the

Encyclopedia of Chemical Technology edited by Kirk and Other (1978,1991ff), which covers the whole range of chemical and associated products. Another encyclopedia covering manufacturing processes is that edited by McKetta(1977). Several books have also been published which give brief summaries of the production processes used for the commercial chemicals and chemical products. The most well known of these is probably Shreve's book on the chemical process industries, now updated by Austin(1984).

The extensive German reference work on industrial processes, *Ullman's Encyclopedia of Industrial Technology*, is now available in an English translation, Ullman(1984).

Specialized texts have been published on some of the more important bulk industrial chemicals, such as that by Miller(1969) on ethylene and its derivatives; there are too numerous to list but should be available in the larger reference libraries and can be found by reference to the library catalogue.

Books quickly become outdated, and many of the processes described are obsolete, or at best obsolescent. More up-to-data descriptions of the processes in current use can be found in the technical journals. The journal *Hydrocarbon Processing* publishes an annual review of petrochemical processes, which was entitled *Petrochemical Developments* and is now called *Petrochemicals Notebook*; this gives flow-diagrams and brief process descriptions of new process developments. Patents are a useful source of information; but it should be remembered that the patentee will try to write the patent in a way that protects his invention, whilst disclosing the least amount of useful information to his competitors. The examples given in a patent to support the claims often give an indication of the process conditions used; through they are frequently examples of laboratory preparations, rather than of the full-scale manufacturing processes. Several short guides have been written to help engineers understand the use of patents for the protection of inventions, and as sources of information.

2. General Sources of Physical Properties

International Critical Tables (1933) is still probably the most comprehensive compilation of physical properties, and is available in most reference libraries. Though it was first published in 1933, physical properties do not change, except in as much as experimental techniques improve, and ICT is still a useful source of engineering data.

Tables and graphs of physical properties are given in many handbooks and textbooks on Chemical Engineering and related subjects. Many of the data given are duplicated from book to book, but the various handbooks do provide quick, easy access to data on the more commonly used substances.

An extensive compilation of thermophysical data has been published by Plenum Press, Touloukian (1970~1977). This multiple-volume work covers conductivity, specific heat, thermal expansion, viscosity and radiative properties.

The Engineering Sciences Data Unit (ESDU) was set up to provide authenticated data for engineering design. Its publication includes some physical property data, and other design data and methods of interest to chemical engineering designers. They also cover data and methods of use in the mechanical design of equipment.

Caution should be exercised when taking data from the literature, as typographical errors often occur. If a value looks doubtful it should be cross-checked in an independent reference, or by estimation.

The values of some properties will be dependent on the method of measurement; for example, surface tension and flash point, and the method used should be checked, by reference to the original paper if necessary, if an accurate is required.

The results of research work on physical properties are reported in the general engineering and scientific literature. The *Journal of Chemical Engineering Data* specializes in publishing physical property data for use in chemical engineering design. A quick search of the literature for data can be made by using the abstracting journals; such as *Chemical Abstracts* （American Chemical Society）and *Engineering Index* (Engineering Index inc., New York).

Computerized physical property data banks have been set up by various organizations to provide a service to the design engineer. They can be incorporated into computer-aided design programs and are increasingly being used to provide reliable, authenticated, design data.

3. Sources of Information on Chemical Engineering

Journals:
(1) Chemical Engineering Science (England)
(2) International Journal of Heat and Mass Transfer (England)
(3) The Chemical Engineering Journal (England)
(4) I and EC—Process Design and Development (America)
(5) I and EC—Product Research and Development (America)
(6) I and EC—Fundamentals(America)
(7) Chemical Engineering Progress(America)
(8) Journal of the American Institute of Chemical Engineers (AIChE Journal)
(9) Chemical Engineering (America)
(10) ChemTech (America)
(11) Environmental and Technology (America)
(12) Journal of Chemical and Engineering Data (America)
(13) Hydrocarbon (America)
(14) Oil and Gas and Petroleum Equipment (America)
(15) Journal of Petroleum Technology (America)
(16) Advances in Heat Transfer (America)
(17) Journal of Applied Polymer Science (America)
(18) The Canadian Journal of Chemical Engineering

Abstract:
(1) Engineering Index
(2) Chemical Abstracts (America)
(3) Theoretical Chemical Engineering Abstracts

Handbooks and Encyclopedia:
(1) Chemical Engineers' Handbook
   R.H.Prrry and C.H.Chilton, 6th edn., McGraw-Hill, 1984
(2) Handbook of Heat Transfer
   W.H.Rohaenow and J.P.Hartnet, McGraw-Hill,1973
(3) Chemical Engineering Practice
   H.W.Gremer and T. Davies, Butterworth，1956～1960
(4) Encyclopedia of Chemical Technology
   Kirk-Othmer, 2th edn, Wiley,1963～
(5) The Materials Handbook

Georges S.Brady,10th edn, McGraw-Hill, 1971

References:

(1) *An Introduction to Industrial Chemistry,* 2nd Edition. C.A.Heaton, Blackie & SonLtd., 1997

(2) Chemical Engineering，Vol 6，2nd Edition, R.K.Sinnott, Pergamon Press, 1996

Selected from "Specialized English for Chemical Engineering and Technology, by Hu Ming etc., Chemical Industry Press, 1998, 236-239"

Words and Expressions

1. International Critical Tables   国际标准数据表。
2. Engineering Sciences Data Unit   工程科学数据组织（英）。
3. flash point   闪点。

Appendix

Appendix 1　化学化工常用构词

1. aci- 酸式
2. aero- 空气
3. –al 醛
4. ald- 醛
5. –aldehyde 醛
6. –amide 酰胺
7. –amine 胺
8. amino- 氨基
9. –ane 烷
10. anhydro- 脱水
11. anti- 反，抗，对，解，阻
12. aryl- 芳（香）基
13. –ase 酶
14. –ate [词尾]用于由词尾为-ic 的酸所组成的盐类或酯类的名称
15. auto- 自，自动
16. benz- 苯基
17. bi- 二，两个，双
18. bio- 生物的
19. bis- 两个，双
20. –carboxylic acid 羧酸
21. chemico- 化学的
22. chemo- 化学
23. chlor- 氯
24. chloro- 氯代，氯（基）Cl—
25. chromato- 色谱
26. chromo- 色
27. cis- 顺式
28. co- 共，同，相互
29. counter- 反，逆
30. cyan- 氰基 CN—
31. cycl(o)- 环（合，化），（循）环
32. de- 脱，去，除，解，减，消，反，止
33. deca- 十，癸
34. dehydro- 脱氢
35. dextro- 右旋的
36. di- 二，双（指基的数目）；联（二）指两个基以一价相连；双（指两个单体相结合）
37. dodeca- 十二
38. electro- 电
39. –en [词尾]指烃或环形化合物
40. endo- 内，桥（环内桥接）
41. –ene 烯
42. epoxy- 环氧
43. –ether 醚
44. ferri- 铁
45. ferro- 亚铁
46. fluo- 氟，荧
47. fluoro- 氟代，氟（基）
48. haem- 血的
49. halo- 卤
50. hepta- 七，庚
51. hetero- 杂，不同
52. hexa- 六，己
53. homo- 同（型），高
54. hydro- 氢化的，氢的，水
55. –ic anhydride 酸酐
56. –ide -化物
57. infra- 在下，较低
58. inter- （在）中（间），互相，合，一起
59. intra- 内
60. iso- 异，同，等
61. –ketone 酮
62. –lactone 内酯
63. laevo- 左旋
64. lipo- 酯的
65. m-(=meta) 间
66. meso- 内消旋，中（间）
67. meta- 间（位）（有机系统名用）；偏（无机酸用）

68.	mono-	一，单		90.	radio-	放射，辐射
69.	multi-	多		91.	re-	再，重，回，向后；相互；相反
70.	nitro-	硝基		92.	retro-	向后
71.	non-	不，非，无		93.	rheo-	流
72.	nona-	九，壬		94.	stereo-	立体，固（体）
73.	o-(=ortho)	邻（位）—		95.	sub-	下，亚，次，副
74.	octa-	八，辛		96.	sulf-	[词头]表示有硫存在
75.	–oic acid	酸		97.	sulfo-	硫代，磺基
76.	–ol	醇，酚		98.	–sulfonic acid	磺酸
77.	–one	酮		99.	super-	过，超，高于
78.	ortho-	正，原，邻（位）		100.	syn-	同，共，与；顺式
79.	–ose	糖		101.	tauto-	互变（异构）
80.	–oside	糖苷		102.	tetra-	四，丁
81.	over-	过（度），超，在外 oxo-氧化，氧代，含氧的		103.	thio-	硫代
				104.	trans-	反（式）；超，跨，过，（以）外，后
82.	para-	（位次）对，仲		105.	tri-	三，丙
83.	penta-	五，戊		106.	ultra-	超，过，（以）外；极端，异常，过度
84.	per-	高，过，全				
85.	phono-	声，音		107.	under-	在下，底下，不足，从属
86.	photo-	光，感光的		108.	uni-	单，一
87.	poly-	多，聚		109.	–yl	（某）基
88.	pre-	预，前，先，在上		110.	–ylene	（某）烯
89.	pyro-	火，热，高温，焦		111.	–yne	（某）炔

Appendix 2　常用有机基团

1.	acetenyl=ethynyl	乙炔基		20.	formyl	甲酰
2.	acetoxy	乙酸，乙酰氧基		21.	heptyl	庚基
3.	acetyl	乙酰基		22.	hexyl	己基
4.	aldo	（表示有醛基存在）醛（元），氧代		23.	hydroxyl(l)	羟基
				24.	methene=methylene	亚甲基
5.	alkoxy	烷氧基		25.	methenyl=methylidyne	次甲基
6.	amino	氨基		26.	methyl	甲基
7.	amyl=pentyl	戊基		27.	naphthyl	萘基
8.	anilino	苯氨基		28.	nitro	硝基
9.	anthraquinonyl	蒽醌基		29.	nitroso	亚硝基
10.	anthryl	蒽基		30.	nonyl	壬基
11.	axo	偶氮基		31.	octyl	辛基
12.	azido	叠氮基		32.	pentyl	戊基
13.	benzoxy=benzoyloxy	苯甲酸基		33.	phenyl	苯基
14.	butyl	丁基		34.	propenyl	丙烯基
15.	carbonyl	羰基		35.	propyl	丙基
16.	carboxy(l)	羧基		36.	sulfo	磺基
17.	decyl	癸基		37.	thio	硫代
18.	diazo	重氮基		38.	vinyl	乙烯基
19.	ethyl	乙基				

Appendix 3 总词汇表

absorber [əb'sɔ:b] n. 吸收器
absorption [əb'cɔ:pʃən] n. 吸收（作用）
absorption [əb'zɔpʃən] n. 吸收，吸收作用
accelerate [æk'seləreit] vt. 加速
accuracy ['ækjurəsi] n. 精确，准确
acetal ['æsitəl] n. 乙缩醛，乙醛缩二乙醇
acetic acid n. 醋（乙）酸
acetone ['æsitəun] n. 丙酮
acetylene [ə'setili:n] n. 乙炔，电石气
acquaint [ə'kweint] vt. 使认识，使熟悉
acquaint oneself with (或 of) 使自己知道（熟悉）
acquaintance [ə'kweintns] n. 熟悉，认识（with）
acrylic [ə'krilik] a. 聚丙烯的，丙烯酸（衍生物）的
activated carbon 活性炭
adhere [əd'hiə] v. 黏附（于），附着（于）；坚持；追随
adiabatic [ædiə'bætik] a. 绝热的，不传热的
adiabatically [ædiə'bætikəli] ad. 绝热地
adsorb [æd'sɔ:b] vt. 吸附
adsorbent [æd'sɔ:bənt] n. 吸附剂
adsorption [æd'sɔpʃən] n. 吸附（作用）
adsorption [æd'sɔ:pʃən] a. 吸附
aerate ['ɛəreit] vt. 充气，鼓气，通风，鼓风
aerosol ['ɛərəsɔl] n. 气溶胶，悬浮微粒
affect [ə'fekt] v. 影响，作用
affect [ə'fekt] vt. 影响
afford [ə'fɔ:d] vt. 担负得起（费用、损失等）；抽得出（时间）；提供
aggregate ['ægrigeit] n. 集合（体），聚集（体）
aggregation [ægri'geiʃən] n. 聚合（作用），聚合体
agitated thin-film evaporator 搅拌式薄膜蒸发器
agitator ['ædʒiteitə] n. 搅拌器，搅拌装置
airlift reaction 升气式反应器
alkaline [ælkəlain] n. 碱性；a. 强碱的
alkane ['ælkein] n. 链烷，烷烃
alkane ['ælkein] n. 烷烃
alkene ['ælki:n] n. 烯烃，链烃
alkyl ['ælkil] n. 烷基
alkylation [ælki'leiʃən] n. 烷基化，烷基取代
alkylbenzene sulfonate 烷基苯磺酸盐
alkyne ['ælkain] n. 炔
alphabetical [ælfə'betikəl] a. 依字母顺序的
alphabetical [ælfə'betikəl] a. 字母顺序的
alum [əlʌm] n. (明、白)矾，硫酸铝
alumina [əlu:minə] n. 矾土，氧化铝

ambient ['æmbiənt] a. 周围的，包围着的
amenable [ə'mi:nəbl] a. 服从的，适合于……的(to)
amine [ə'mi:n] n. 胺
ammoniated [ə'məunieitid] a. 充氨的，含氨的
ammonium carbamate 氨基甲酸铵
amorphous [ə'mɔ:fəs] a. 无定形的，非晶体的
an educated guess 有根据的推测
analyte ['ænəlait] n. 被分析物
analyze ['ænəlaiz] vt. 分析
analyze ['ænəlaiz] vt. 分析；分解
aniline ['ænə,lin] n. 苯胺
anionic [æno'ɔnik] a. 阴离子的
anode ['ænəud] n. 阳极，正极
Antarctic [ænt'ɑ:ktik] n.;a. 南极地带(的)
anthracite ['ænθrəsait] n. 无烟煤；硬煤
antiknock ['ænti'nɔk] a. 抗爆的，防爆的，抗震的
apiezon [ə'pi:zɔn] n. 阿匹松，饱和烃
apparatus [æpə'reitəs] n. 设备
appearance [ə'piərəns] n. 外貌，外表，出现，露面
approach [ə'prəutʃ] n. 途径，方法，手段
argon ['a:gɔn] n. 氩
argument ['a:gjumənt] n. 争论，论证
aromatization n. 芳构化
as a consequence of 由于……（结果）
as long as … 只要……
ascension [ə'senʃən] n. 上升，升高
ascertain [æsə'tein] vt. 弄清，确定
asphalt ['æsfælt] n. (地)沥青，柏油
asphalt base crude 沥青基石油
assembly [ə'sembli] n. 装配，组合
attorney [ə't3:ni] n. 代理人，律师
authenticity [ɔ:θen'tisiti] n. 可靠性，真实性
automate ['ɔ:təmeit] v. 使……自动化
aviation [eivi'eiʃn] n. 航空，飞行
awe [ɔ:] n.;v. (使)敬畏（畏惧）
azeotrope [ə'zi:ətrəup] n. 恒沸物，共沸混合物
backbone ['bækbəun] n. 构架，骨干，主要成分
back-mixing 返混
barren ['bærən] a. 不毛的，贫瘠的
be + 不定式或其短语 拟将……，准备……
be based on… 以……为依据；基于
be described as 说成是，被称作
be divided into… 被分为……
be formed from… 由……生成，由运输，输送，运送形成

be interfaced with　与……连接
be removed to　移动
be subject to　易受
be thought of…　被认为是……
be under control　受到控制；受到支配
behavior [bi'heiviə]　n. 性能，性质，行为
bench [bentʃ]　n. 实验台，装置
benefit ['benifit]　vt.有益于，vi. 受益，n. 利益，好处
benign [bi'nain]　a. 有益于健康的,良好的,[医]良性的
benzenoid　a. 苯（环）形的
beset [bi'set]　vt. 包围，缠绕，为……所苦
bestow [be'stəu]　vt. 把……赠予（给）
bicarbonate [bai'kɑ:bənit]　n. 碳酸氢盐,酸式碳酸盐
bifunctional　a. 双官能团的
biodegradable　a. 可生物降解的
biopolymer　n. 生物聚合物
bisphenol A　双酚A
bitumen [bə'tu:mən]　n. 沥青，柏油
bituminous [bi'tju:minəs]　adj. 含沥青的
blast furnace　高炉，鼓风炉
blast furnace iron-making and foundry　高温炼铁和铸造
bleach ['bli:tʃ]　v.漂白
blueprint [blu:print]　n. 蓝图，设计图
boiling point　沸点
boil-up　蒸出（蒸汽）
bollworm　n. 螟蛉
bombardment [bɔm'bɑ:dmənt]　n. 照射,辐照;轰击，打击
bone-dry　a. 干透的
bottom ['bɔtəm]　n.(pl.) 底部沉积物，残留物，残渣
boundary layer　边界层
branched　a. 有枝的，分岔的
break down　毁掉，制服，压倒，停顿，倒塌，中止，垮掉，分解
breakthrough ['breikθru:]　n. 穿过，透过，突破
britle ['britl]　a. 脆的，易碎的
BTU　英制热量单位
Btu=British thermal unit　英热量单位（=252卡）
bubble ['bʌbl]　n. 气泡；v.(使)起泡
bubble column　鼓泡塔
bubble-cap tower　泡罩塔
buffer ['bʌfə]　n. 缓冲液，缓冲器
building block　结构单元;预制件;积木
bulk [bʌlk]　n. 块，主体
bulk density　堆积密度
bulk polymerization　本体聚合
bumper ['bʌmpə]　n. 保险杠

buoyancy ['bɔiənsi]　n. 浮力，浮性
buoyant ['bɔiənt]　a. 有浮力的，能浮的，易浮的
butane ['bju:tein]　n. 丁烷
butene ['bju:ti:n]　n. 丁烯
by convention　按照惯例
by the time　当……时
byproduct ['baiprɔdʌkt]　n. 副产物
cabinet ['kæbinit]　n. 箱，室，壳体
calandria [kə'lændriə]　n. 排管式，加热管群
calcine ['kælsin]　v.; n. 煅烧，烧成（灰）
calcium carbide　电石
calibrate ['kælibreit]　vt. 校准
capillary gel eletrophoresis　毛细管凝胶电泳
capillary isoelectric focusing　毛细管等电聚焦
capillary electrochromatography　毛细管电色谱
capillary electrophoresis　毛电管电泳
capillary zone electrophoresis　毛细管区带电泳
carbamate ['kɑ:bəmeit]　n. 氨基甲酸酯
carbon black　炭黑
carbonate ['kɑ:bəneit]　n. 碳酸盐
carbonium　n. 阳碳，正碳
carbonium ion mechanism　正碳离子机理
carbonylation　n. 羰化作用
carpet tile backing　n. 地毯背衬
carrier gas ['kæriə gæs]　n. 载气
cascade [kæs'keid]　v.; n. 梯流，阶流式布置；级联，串联，阶式
casein ['keisiin]　n. 酪素，酪蛋白
casing-head gas　油井气，油田气
catastrophic [kætə'strɔfik]　a. 灾难性的，灾变的
cathedral [kə'θi:drəl]　n. （一个教区内的）总教堂，大教堂；a. （像）大教堂的，权威的
cathode ['kæθəud]　n. 阴极，负极
cationic　a. 阳离子的
cease ['si:s]　v. 停止，终止
celite [si'lait]　n. 硅藻土
celluloid ['seljuloid]　n. 赛璐珞，硝酸纤维素
centrifugal [sen'trifjugəl]　a. 离心的，离心力的，利用离心力的
centrifuge ['sentrifjudʒ]　n.;v. 离心，离心机，离心器
ceramics [si'ræmiks]　n. 陶瓷制品，陶瓷学
certitude ['sə:titju:d]　n. 确实，必然性
chalk [tʃɔ:k]　n. 白垩，方解石
chamber ['tʃeimbə]　v. 室，房间
channel ['tʃænl]　v. 沟流；n. 沟流槽
chelation [ki'leiʃən]　n. 螯合作用
chemical vapor deposition　化学气相淀积
chlorate ['klɔ:rit]　n. 氯酸盐

chlorofluorocarbon n. 含氯氟烃(CFC)
chloroform ['klɔrəfɔ:m] n. 三氯甲烷，氯仿
chromate ['krəumeit] n. 铬酸盐
chromatogram ['krəumətəgræm] n. 色谱图
chromatography [,krəumə'tɔgrəfi] n. 色谱法，层析法，层离法，色层分析法
circulating liquor crystallizer 母液循环结晶器
circulating magma crystallizer 晶浆循环结晶器
circumvent [sə:kəm'vent] vt. 绕过，回避，胜过
clean [kli:n] a. 彻底的，完全的
coagulant [kəu'ægjulənt] n. 混凝剂
coagulation [kəu,ægju'leiʃən] n. 凝结，絮（胶）凝，聚沉
coalescence [,kəuə'lesəns] n. 聚结
coarse [kɔ:s] a. 粗糙的，粗鄙的
coal tar 煤焦油；煤黑油
coax [kəuks] vt. 耐心地处理,慢慢地把……弄好；诱,哄
cobalt ['kəubɔ:lt] n. 钴（Co）
coefficient [,kəui'fiʃənt] n. 系数，率
coke oven gas 焦炉煤气
collect [kə'lekt] vt. 收集
colligative ['kɔligətiv] a. 依数的
colloid ['kɔlɔid] n. 胶体；adj. 胶体的
colloid mill 胶体磨
columblum [ke'lʌmbiəm] n. 铌
column ['kɔləm] n. 柱，支柱，圆柱
combination of coal steam gasification and methanation 煤的水蒸气气化与甲烷化结合
combustible [kəm'bʌstəbl] adj.; n. 易燃的;易燃物
commencement [kə'mensmənt] n. 开始，开端，开工
commitment [kə'mitmənt] n. 承担义务
compartmentalization [kɔmpɑ:tmentəlai'zeiʃən] n. 区域化，隔开
compatible [kəm'pætəbl] a. 兼容的,可共存的;一致的,相似的,协调的
competence ['kɔmpitəns] n. 能力
composite ['kɔmpəzit] n. 复合材料，合成[复合，组合]物
composting ['ka:mpəustiŋ] n. 堆制肥料
concept ['kɔnsept] n. 概念，观念，思想
concise [kən'sais] a. 简明的，扼要的，短的
concurrent [kən'kʌrənt] a. 并流的，顺流的
condensation [kɔnden'seiʃən] n. 凝结，凝固，浓缩
condense [kən'dens] v. 冷凝，缩合
condenser [kən'densə] n. 冷凝器
conduct ['kɔndəkt] n. 行为

conduction [kən'dʌkʃən] n. 传导（性，率），导热性，导电性，热导率
conductivity [kəndʌk'tiviti] n. 传导率，热导率
conduit ['kɔndit] n. 导管，输送管，（大）管道
confirm [kən'fə:m] vt. 进一步证实，确定
connotation [,kɔnə'teiʃən] n. 含意
consensus [kən'sensəs] n. 一致
constant ['kɔnstənt] n. 常数
constituent [kəns'titjuənt] n. 组成，组分，成分
constitutional [,kɔnsti'tju:ʃənəl] a. 构成的
constraint [kən'streint] n.强制，制约，限制
contaminant [kən'tæminənt] n. 污染物
contaminant [kən'tæminənt] n. 沾污物，污染物
contaminate [kən'tæmineit] vt. 污染，弄脏，毒害
contamination [kəntæmi'neiʃən] n. 污染，污物
continuous circulation band dryer 连续循环带式干燥器
control [kən'trəul] n. 对照（物），控制
convection [kən'vekʃən] n.（热，电）对流，迁移
converge [kən'və:dʒ] v. 会聚，汇合，【数】收敛
conversion [kən'və:ʃən] n. 转化，转换
conveyor dryer 带式干燥器
cook ['kuk] vt. 烹调
copolymer [kəu'pɔlimə] n. 共聚物
copperas ['kɔpərəs] n. 绿矾，硫酸亚铁
correlation [kɔri'leiʃən] n. 相互关系，伴随关系，关联（作用）
corrosive [kə'rəusiv] a. 腐蚀的
corrosiveness n. 腐蚀性，侵蚀作用
countercurrent ['kauntə'kʌrənt] a. 逆流的，对流的
counterpart ['kauntəpɑ:t] n. 对应物，配对物，对方；一对东西中之一，副本
craft [kræft] n. 手艺，技艺
critical velocity 临界速度
cross section （横）截面，剖面，断面
cross-current 错流，正交流
cross-link ['krɔ:sliŋk] v.（聚合物）交联，横向耦合
crucial ['kru:ʃəl] a. 决定性的，重要的
crude benzene n. 粗苯
crystal size distribution （缩写为CSD)晶体大小的分配
crystalline ['kristəlain] a. 结晶的，结晶状的；水晶的，结晶（质）的；结晶状的
crystallization [kristəlai'zeiʃən] n. 结晶（作用，过程）
crystallize ['kristəlaiz] vt.;vi.（使）结晶
crystalloid ['kristə,lɔid] a. 结晶的；n. 晶体
CSTR=continuously stirred tank reactor 连续搅拌釜反应器

customer ['kʌstəmə] n. 顾客，主顾，用户
cyanide ['saiə,naid] n. 氰化物
cycloalkane ['saikləu'ælkein] n. 环烷
cyclobutane [,saikləu'bju:tein] n. 环丁烷
cyclohexanol [saikləu'heksənɔl] n. 环己醇
cyclohexanone [saikləu'heksənəun] n. 环己酮
cyclopentane [,saiklə'pentein] n. 环戊烷
cyclopropane [,saikləu'prəupein] n. 环丙烷
datum ['deitəm] (复 data['deitə]) n. 资料，论据
deactivate [di:'æktiveit] vt. 减活，去活化，钝化
decelerate [di:'seləreit] v. 减速，减慢
decomposition [di:kɔmpə'ziʃən] n. 分解，离解
definiteness ['definitnis] n. 明确，确定，肯定
degree [di'gri:] n. 程度，度
degree of polymerization 聚合度
dehydrogenation [di:,haidrədʒə'neiʃən] n. 脱氢（作用）
deliberately [di'libərətli] ad. 故意地，蓄意地；审慎地，深思熟虑地
delineation [dilini'eiʃən] n. 描述，叙述
demethylation n. 脱甲烷（作用）
dependent upon (depend on) 依靠的，依赖的，由……决定的，随……而定的
deployment ['diplɔimənt] n. 使用，利用，推广应用
derivative [di'rivətiv] n. 衍生物
derivatization n. 衍生法
desalination [di:,sæli'neiʃən] n. 脱盐
desalt ['di:'sɔ:lt] vt. 脱盐
desirable [di'zairəbl] a. 合意的，希望得到的，合乎需要的
desorption [di:sɔ:pʃne] n. 解吸作用
destabilize [di:'steibilaiz] vt. 使脱稳
desulphurization [di:sʌlfərai'zeiʃən] n. 脱硫，除硫
detached [di'tætʃt] a. 分离的，孤立的
detector [di'tektə] n. 检测器
detector signal [signl] 检测器信号
determination [ditə:mi'neiʃən] n. 测定，确定
detrimental [detri'mentl] a. 有害的，不利的
devastate ['devəsteit] vt. 破坏，毁坏，使荒芜
devolatilization n. 脱（去）挥发分（作用）
dialkyl [dai'ælkil] phthalate n. 邻苯二甲酸二烷基酯
dialysis [dai'ælisis] n. 渗透，透析
diammonium hydrogen phosphate 磷酸氢二铵
diamond [,daiəmənd] n. 金刚石，钻石
diaphragm [,daiəfræm] n. 隔膜，隔板
diene [daii:n] n. 二烯（烃）
diesel ['di:zl] n. 内燃机，柴油机

diesel oil 柴油
diet ['daiət] n. 食物,饮食
differentiate [difə'renʃieit] v. 区分，区别；求微分，求导数
diffuse [di'fju:z] v. 扩散，散布
diffusivity [difju:'siviti] n. 扩散性，扩散系数
digestive [di'dʒestiv] a. 消化的,助消化的
digitize ['didʒitaiz] vt. 将数字化，数字化处理
dimension [di'menʃən] vt.尺寸
dimensionless [di'menʃənlis] a. 无因次的，无量纲的
dimerization [daimərai'zeiʃən] n. 二聚（作用）
diminish [di'miniʃ] v. 减少，递减，削弱，由大变小
dinitrogen [dai'naitrədʒən] n. 分子氮，二氮
dipole ['daipoul] n. 偶极（子）
direct heated evaporator 直接加热蒸发器
disinfectant [disin'fektənt] n. 消毒剂，杀菌剂
disperse [dis'pə:s] v. （使）分散，弥散
disperser [dis'pə:sə] n. 分散器，泡罩
display [dis'plei] vt；n. 显示，表现
dissipate ['disipeit] v. 使耗散，消除，消耗
dissipated ['disipeitid] a. 散逸的 浪费的
dissipation [disi'peiʃən] n. 耗散，损耗，消散
distil(l) [dis'til] vt. 蒸馏，用蒸馏法提取；提取……的精华
distinction [dis'tiŋkʃən] n. 区别，差别，特性
distinguish [dis'tiŋgwiʃ] vt. 区别
divert [dai'və:t] vt. 使转向，使变换方向，转移
dosage ['dəusidʒ] n. 用（剂）量
draftsman=draughtsman ['dra:ftsmən] n. 制图员，绘图员，起草者
drying ['draiiŋ] n. 干燥
dumped packing 乱堆填料
duplicate ['dju:plikeit] vt. 重复，加倍，复制
dynamic [dai'næmik] a. 动力（学）的
eddy ['edi] n. （水，风，气等的）涡，旋涡
effect [i'fekt] vt. 实现；n. 效应，影响，效果
effluent ['efluənt] n. 流出物；a. 流出的
electrochemical [i'lektrə'kemikəl] a. 电化学的
electrolyse [i'lektrəlaiz] vt. 电解（=electrolyze）
electrolysis [ilek'trɔlisis] n. 电解法，电解作用，电分析
electrolyte [i'lektrəulait] n.电解质
electrolytic [i'lektrəu'litik] a. 电解的，电解质的
electroosmotic [i,lektrəuɔz'məutik] a. 电渗的
electrophoresis [i,lektrəfə'ri:sis] n. 电泳
eliminate [i'limineit] n. 除去，排除，消除
elude [i'lu:d] vt. 使困惑，难倒
eluent ['eljuənt] n. 洗提液

embed [im'bed] *vt.* 把……嵌入，栽种
embodiment [im'bɔdimənt] *n.* 具体化，具体体现
emerge [i'mə:dʒ] *vi.* 出现，形成
emergence [i'mə:dʒəns] *n.* 出现，浮现
empirical [em'pirikəl] *a.* 经验（上）的
emulsion [i'mʌlʃən] *n.* 乳状液
emulsion polymerization 乳液聚合
encase [in'keis] *vt.* 把……放入套内
encounter [in'kauntə] *vt.* 遇到
encounter [in'kauntə] *v.* 遇见，遭遇，冲突
endothermic [endəu'θəmik] *a.* 吸热的
enthalpy [en'θælpi] *n.* 焓，热焓，（单位质量的）热含量
entity ['entiti] *n.* 存在，实体，统一体
entropy ['entrəpi] *n.* 熵
enunciate [i'nʌnsieit] *v.* 明确叙述
envision [in'viʒən] *vt.* 想象，预想
epitaxy [epi'tæksi] *n.* 晶体取向生长
equilibrium [i:kwi'libriəm] *n.* 平衡
equilibrium state 平衡状态
erratically *a.* 不规律的，不稳定的
establish [is'tæbliʃ] *vt.* 建立
ethane ['eθein] *n.* 乙烷
ethanolamine ['eθə'nɔlemi:n] *n.* 乙醇胺
ether ['i:θə] *n.* 醚
ethyl ['eθil] *n.* 乙基
ethylbenzene *n.* 乙（基）苯
evaporating [i'væpə,reitiŋ] *a.* 蒸发用的，蒸发作用的
exhaust [ig'zɔ:st] *n.* 排出，排气；*vt.* 用尽，排出
exhaust gas 废气
exothermic [eksəu'θə:mik] *a.* 放热的
exploration [exsplɔ:'reiʃən] *n.* 探究，探索，探测
exponent [eks'pəunənt] *n.* 指数，幂（数），阶
extensive [iks'tensiv] *a.* 广泛的
extraction [ik'strækʃən] *n.* 萃取
extractive metallurgical 湿法冶金的
extrapolation [ekstræpə'leiʃən] *n.* 外推法，推断，推知
extruder [eks'tru:də] *n.* 挤压机，（螺旋）压出机
fabricate ['fæbrikeit] *vt.* 制造，生产，制备；装配，安装，组合
facility [fə'siliti] *n.* 工厂，研究所，实验室，装备
factor ['fæktə] *n.* 因素；要素；因子
fashion ['fæʃən] *n.* 型，样式
fatty acid ['fæti'æsid] 脂肪酸
feasibility [fi:zə'biliti] *n.* 可行，可实行
feature ['fi:tʃə] *n.* 要点，特征
feed [fi:d] *n.* 进料，加料；加工原料

feeder ['fi:də] *n.* 给水管，进料器
feedstock [fi:d'stɔk] *n.* 原料
fender ['fendə] *n.* 挡泥板，防护板
ferric ['ferik] *a.* (三价) 铁的
ferrous ['ferəs] *a.* (二价) 亚铁的
filament ['filəmənt] *n.* （细）丝，（细）线；灯丝，游丝
fill [fil] *vt.* 填充，装满
filter ['filtə] *n.* 过滤器，滤光器，筛选；*vt.* 过滤，渗透；*vi.* 滤过，渗入
filtration [fil'treiʃən] *n.* 过滤
final ['fainl] *a.* 最后的；最终的
a final product 最终产品
finishing ['finiʃiŋ] *n.* 精加工，最终加工
finite ['fainait] *a.* 有限的，受限制的
firebrick ['faiəbrik] *n.* 耐火砖
flare [flɛə] *v.* 端部张开，（向外）扩张（成喇叭形）
flash dryer 气流干燥器
flask [flɑ:sk] *n.* 烧瓶，长颈瓶
flocculation [flɔkju'leiʃən] *n.* 絮凝，絮接产物
fluctuation [flʌktju'eiʃən] *n.* 脉动，波动，起伏，增减
flue [flu:] *n.* 烟道，风道
fluffy ['flʌfi] *a.* 蓬松的，松软的
fluidized bed dryer 流化床干燥器
fluorinated ['fluərineitid] *a.* 氟化的
fluorine ['fluəri:n] *n.* 氟（F）
flux [flʌks] *n.* 通量
foodstuff ['fu:dstʌf] *n.* 食品，粮食
Formica [fɔ:'maik] *n.* 佛米卡（一种家具表面抗热塑料贴面）
forward reaction 正反应
fossil ['fɔsil] *n.* 化石
foul [faul] *vt.* 弄脏，淤塞；*vi.* 腐烂，缠结；*a.* 淤塞的
fouling [fauliŋ] *n.* 污垢
foundry ['faundri] *n.* 铸造
fractionate ['frækʃəneit] *vt.* 使分馏，把……分成几部分
Frash process 地下熔融法
free flowing 自由流动的，流动性能良好的
free radical mechanism 自由基机理
free solution CE 自由溶液毛细管电泳
frigid ['fridʒid] *a.* 寒冷的，严寒的
furan ['fjuəræn] *n.* 呋喃
fuse [fju:z] *v.* 熔融，熔化
gas chromatography 气体色谱（法）
gas oil 粗柴油，瓦斯油，汽油

gas-liquid partition chromatography　气液分配色谱法
gauze [gɔ:z]　n.（金属丝，纱，线）网
gazette [gə'zet]　n. 公报，报纸
gelatin ['dʒelətin]　n. 凝胶，白明胶
generation ['dʒenə'reiʃən]　n. 产生，形成，引起
generic [dʒi'nerik]　a. 属的，类的，一般的，普通的
glycerolysis of fats　脂肪甘油水解
glycol ['glaikɔl]　n. 乙二醇
(be) good for　对……适用，有效，有利，有好处
grandeur ['grændʒə]　n. 宏伟，壮观；伟大，崇高；富丽堂皇，豪华
granular ['grænjulə]　a. 颗粒状的，晶状的
granulate ['grænjuleit]　vt. 使成颗粒，使成粒状
greenhouse ['gri:nhaus]　n. 温室，暖房
growth [grəu]　v. 成长，生长；长大
hafnium ['hæfniəm]　n. 铪
harness ['ha:nis]　vt. 控制，治理，利用，开发
have acquaintance with…　熟悉，认识
hazard ['hæzəd]　n. 危害
H-bond　n. 氢键
heavy distillate　重馏分
heavy flotation oil　重质浮选油
heavy mineral oil　重质矿物油
heherogeneous ['hetərəu'dʒi:njəs]　a. 多相的，非均匀的
heretofore ['hiətu'fɔ:]　ad. 至今，到现在为止；在此以前
heritage ['heritidʒ]　n. 遗产，继承物
heterogeneous [,hetərə'dʒi:niəs]　a. 非均相的，不均的
high heat valve　热值高
HMDS (Hexamethyl Disilazane)　n. 六甲基二硅氮烷
hold-down　n. 压具（板，块），压紧（装置），固定
holdup ['həuldʌp]　n. 容纳量，塔储量
hybrid ['haibrid]　a. 杂交的，混合的
hybridization [haibridai'zeiʃən]　n. 杂交，杂化
hydrate ['haidreit]　v. （使）水合（化）（使成水合物）
hydrochloric ['haidrə'klɔrik]　a. 氯化氢的
hydrochloric acid　盐酸
hydrocracking　n. 加氢裂化，氢化裂解
hydrodementallize　v. 加氢脱金属
hydrodesulfurization　加氢脱硫过程
hydrodynamics ['haidrəudai'næmiks]　n. 流体动力学
hydrofluoric [haidrəuflu'ɔrik]　a. 氟化氢的，氢氟酸的
hydrogenate [hai'drɔdʒəneit]　vt. 使与氢化合，使氢化
hydrogenation [haidrɔdʒi'neiʃen]　n. 加氢（作用）
hydrophilic [haidrə'filik]　a. 亲水的

hydroprocess　加氢过程
hydroxide　n. 氢氧化物
hypochlorite [haipə'klɔ:rait]　n. 次氯酸盐
hypochlorous acid　次氯酸
hypothetical [,haipə'θetikl]　a. 假设的，假定的，爱猜想的，有前提的
identical [ai'dentikəl]　a. 相同的，完全相同的
identification [aidentifi'keiʃən]　n. 鉴别，鉴定
immerse [i'mə:s]　vt. 浸（入，没），沉入；专心，埋头于，投入
immiscible [i'misibl]　a. 不（能）混合的，非互溶的
immobile [i'məubail]　a. 不能移动的，固定的，静止的
immobilization [i'məubilai'zeiʃən]　n. 固定,定位，降低流动性
immune [i'mju:n]　a. 免除的，可避免的；不受影响的
impervious [im'pə:viəs]　a. 不能透过的,不可渗透的
impetus ['impitəs]　n. (推)动力，促进，动量，冲量
implement ['impliment]　n. 工具，器具
implementation [implimen'teiʃən]　n. 实现，实施，履行
implicitly [im'plicitli]　ad. 含蓄地；无疑地，无保留地，绝对地
impose [im'pəuz]　① 征税；② 把……强加给……
impregnate ['impregneit]　vt. 浸渍
impure [im'pjuə]　a. 不纯的，有杂质的
in a practical sense　实际上
ingeo [in'dʒiəu]　n. 聚乳酸纤维
in such a manner that……　以这样的方式以致……
in view of…　鉴于……，考虑到……
incineration [insinə'reiʃən]　n. 焚化，灰化，煅烧
incredible [in'kredəbl]　a. 难以置信的，不可思议的，惊人的
individual [,indi'vidjuəl]　a. 个人的，个别的，单独的
inductive [in'dʌktiv]　a. 引入的，导论的，诱导的，归纳的
inert [i'nə:t]　n. 惰性组分；a. 惰性的，不活泼的，惯性的
infrared ['infrə'red]　a. 红外线的
infringement [in'frindʒmənt]　n. 侵害，违反
ingredient [in'gri:diənt]　n. 组成部分，成分
inject [in'dʒekt]　vt. 注射
injector [in'dʒektə]　n. 注射器
insect ['insekt]　n. 昆虫
insecticide [in'sektisaid]　n. 杀虫剂,农药
insight　了解
insight ['insait]　n. 洞悉，洞察
install [in'stɔ:l]　vt. 安装，安置

integrity [in'tegriti]   n. 完整，完全，完善
interconnection [intə(:)kə'nekʃən]   n. 互相联系，互连
intercontinental [,intəkɔnti'nentl]   a. 洲际的
interdependent [intədi'pendənt]   a. 互相依赖的，互相影响的
interfere with   干涉，干扰，妨碍
interficial polymerization   界面聚合
intermediate distillate   中间馏分
interpolation [intɜ:pə'leiʃən]   n. 内插法，插入，解释
intimately ['intimitli]   ad. 密切地，紧密的
irrotational [irəu'teiʃənl]   a. 无旋的，不旋转的
isobutene [aisəu'bju:ti:n]   n. 异丁烯
isobutane [,aisəu'bju:tein]   n. 异丁烷
isocyanate ['aisəu'saiəneit]   n. 异氰酸盐（酯）
isomerization   n. 异构化（作用）
isopentane [aisəu'pentein]   n. 异戊烷，2-甲基戊烷
isoprene ['aisəupri:n]   n. 异戊二烯
isothermal [aisəu'θɜ:məl]   a.; n. 等温（的），等温线（的）
isothermal [,aisou'θə:ml]   a. 等温线的
issue ['isju:]   n. 问题，论点，流出物，流出
it is common practice (+ inf.)   通常的做法是……
jet condenser   喷射式冷凝器
jet fuel   喷气式发动机燃料
ketone ['ki:təun]   n. (甲)酮
kiln [kiln, kil]   n. 窑，炉；v. 窑烧
kilowatt ['kiləwɔt]   n. 千瓦（特）
kineticist [kai'netisist]   n. 动力学家
kinetics [kai'netiks]   n.动力学
labile ['leibail]   a. 易发生变化的，易分解的，不稳定的
laminar ['læminə]   a. 层流的，层状的
lapse [læps]   n.;v. 失效，终止
latent heat   潜热
lateral ['lætərəl]   a. 横向的，水平的
latex ['leiteks]   n. 橡胶，乳状物，（天然橡胶，人造橡胶）乳液
lay [lei]   a. 外行的，局外的
layout ['leiaut]   n. 草图，布置图，布置，安排
lead chamber process   铅室法
leftover   a.;n. 剩余的(物)
leguminous [le'gju:minəs]   a. 豆科的，似豆科植物的
lengthy ['leŋθi]   adj. 漫长的
liability [laiə'biliti]   n.责任，义务
light distillate   轻馏分
light heating oil   轻质燃料油

lignin ['lignin]   n. 木质素
lignite ['lignait]   n. 褐煤
linear ['liniə]   a. （直）线的，直线形的，线性的，线性化的
lipids ['lipidz]   n. 脂类
liquefaction [,likwi'fækʃən]   n. 液化
locant ['ləukənt]   n. 位次
loop [lu:p]   n. 环，回路，循环;vt. 使成环，以环连接；vi. 打环
loop reactor   环路反应器
LPG=liquefied petroleum gas   液化石油气
lubricating oil   润滑油
lye [lai]   n. 碱液
lyophilic [,laiə'filik]   a. 亲液的
lyophobic [,laiə'fɔbik]   a. 疏液的，憎液的
macromer   n.高聚物
macroscopic [,mækrəu'skɔpik]   a. 宏观的，肉眼可见的
magma ['mægmə]   n. 稀糊状混合物，岩浆，稠液
magnitude ['mægnitju:d]   n. （数）量级，大小
magnitude ['mægnitju:d]   n. 巨大，重大
mainstream ['meinstri:m]   n. 干流，主流；主要倾向
major ['meidʒə]   a. 主要的
maldistribution ['mældistri'bju:ʃən]   v.; n. 分布不均，分布不当
manifold ['mænifəuld]   n. 总管，集气管，导管
manipulate [mə'nipjuleit]   vt. 熟练的操作，使用机器的，利用
manufacture [mænju'fæktʃə]   n. 产品，制造
market ['mɑ:kit]   n. 市场，菜市场；vt. （在市场上）销售
mass ['mæs]   n. 质量
mass separating agent   质量分离剂
medicinal [me'disinl]   a. 药的，药用的；n. 药物，药品
melamine ['meləmi:n]   n. 蜜胺，三聚氰（酰）胺
membrane ['membrein]   n. 膜，薄膜，隔板
mercury ['mə:kjuri]   n. 汞，水银（Hg）
merge [mə:dʒ]   v. 合并，汇合
metallurgy [me'tælədʒi]   n. 冶金(学)
metaphor ['metəfə]   n. 隐喻，比喻
metastable [,metə'steibl]   a. 亚稳的
metathesis [me'tæθəsis]   n. 复分解（作用），置换（作用）
methanation [,meθə'neiʃən]   n. 甲烷化作用
methane ['meθein]   n. 甲烷，沼气
methyl ester   甲酯
methyl ['meθil]   n. 甲基

methylchloroform n. 三氯乙烷，甲基氯仿
methylcyclopentane n. 甲基环戊烷
micellar electrokinetic capillary chromatography 胶束电动毛细管色谱
micelle [mi'sel] n. 胶束 胶囊
microbe ['maikrəub] n. 微生物,细菌
microscope ['maikrəskəup] n. 显微镜
microscopic [,maikrəs'kɔpik] a. 微观的，微小的，显微镜的
millennium [mi'leniəm] n. 一千年
mimic ['mimk] a.; vt.; n. 模仿的，假装的，仿造物（品）
mind-numbing 失去感觉的心情（头脑），迟钝的感觉 神志（头脑）
mineral ['minərəl] n. 矿物质
minor ['mainə] 较小的，较次要的
minute ['minit] a. 微小的
mixed base crude 混合基石油
mixed-phase reactor 多相反应器
module ['mɔdju:l] n. 组件
modulus ['mɔdjuləs] n. 模数，系数，指数
molecular beam 分子束
molecular beam generated deposit 分子束淀积
molecular structure 分子结构
molybdenum [mɔ'libdinəm] n. 钼（Mo）
motor gasoline 动力汽油，车用汽油
moving trickle bed reactor 移动滴流床反应器
multiple-feed 多口进料
multitude [mʌltitju:d] v.; n. 众多，大量
municipal [mju:'nisipl] a. 市政的，城市的
naphthene base crude 环烷基石油
nasty ['nɑ:sti] a. 难处理的，极坏的，（气味）令人作呕的
negligible ['neglidʒəbl] a. 可忽略的，不计的，很小的
neopentane [,ni:əu'pentein] n. 新戊烷
network ['netwə:k] n. 网络
nickel ['nikl] n. 镍（Ni）
nitration [nai'treiʃən] n. 硝化（作用），渗氮（法）
nitrile ['naitrail] n. 腈
nitro- [词头] 硝基
nitrogenous [nai'trɔdʒinəs] a. 含氮的
nitroglycerine [naitrəu'glisəri:n] n. 硝化甘油（炸药），硝酸甘油
nomenclature [nə'menklətʃər] n. 命名法，术语
nonaggregate [,nɔn'ægrigit] v. 不聚集，不聚结成团
noncake ['nɔn'keik] v. 不结成块，使……不结块
non-destructive a. 无破坏性的，无害的

nonetheless [nʌnðə'les] ad. 仍然，不过（=nevertheless）
not…, …nor 不……，也不……
noxious ['nɔkʃəs] a. 有毒的，有害的，不卫生的
nucleic acid 核酸
nullify ['nʌlifai] v. 废除，使……无效
numerator ['nju:məreitə] n. (分数中的)分子
numerical [nju:'merikəl] a. 数（量，字，值）的，用数字表示的
obliterate [ə'blitəreit] vt. 除去，删去，消除
occasionally [ə'keiʒənəli] ad. 偶尔，有时，间或
olefin ['əuləfin] n. 烯烃
oleum n. 发烟硫酸
oligomer n. 低聚物，（分子量）聚（合）物
one-pass 单程，非循环过程
onset ['ɔnset] n. （有力的）开始，发动
on-site a. (在)现场的,就地的
opacity [əu'pæsəti] n. 不透明性
opaque [əu'peik] a. 不透明的，不传导的，无光泽的
open-cast a.; ad. 露天开采的（地）
opposing [ə'pəuziŋ] n. 相反，相对，反抗
opt [ɔpt] vi. 选择，挑选(for, between)
order ['ɔ:də] n. 级数
ordered packing 整砌填料，规整填料
orthokinetic [ɔ:θəkai'netik] a. 同向移动的
outside heated steam gasification 外热式水蒸气气化
oven ['ʌvn] n. 炉，烘箱
over-all 总的，全部的
ozonation n. 臭氧化作用
ozone ['əuzəun] n. 臭氧
pack [pæk] vt. 填充 包装 压紧
package ['pækidʒ] n. 包装，打包；（中、小型的）包裹，包；（商品、产品等的）一件
packing ['pækiŋ] n. 填料，填充物，包装，打包
panoramic [pænə'ræmik] a. 全景的，全貌的
paradoxically [pærə'dɔksikəli] ad. 似非而可能是，自相矛盾地，荒谬地
paraffin base crude 石蜡基石油
parameter [pə'ræmitə] n. 参数，系数
paramount ['pærəmaunt] a. 首要的，最高的
partial pressure ['pɑ:ʃəl'preʃə] 分压
partition [pɑ:'tiʃən] n.; vt 分离，分开，分配
patent ['peitənt] n. 专利，专利权, vt. 取得……的专利权
patentee [pætən'ti] n. 专利权所有人
pathway ['pɑ:θwei] n.路径，通道，轨迹
peat [pi:t] n. 泥煤，泥炭

peculiar [pi'kju:ljə] a. 特有的，独特的，特殊的；奇怪的
pedestrian [pi'destriən] a. 普通的，平凡的；n. 非专业人员
pelletize ['pelitaiz] v. 造粒，做成丸（球，片）状
penalty ['penlti] n. 罚款，损失
pentane ['pentein] n. 戊烷
peptization [,peptai'zeiʃən] n. 胶溶作用，解胶（作用），塑解
perikinetic [,perikai'netik] a. 异向的，与布朗运动有关的
peroxide [pə'rɔksaid] n. 过氧化物
perpendicularly [pə:pən'dikjuləli] ad. 垂直地，正交地，直立地
pertaining [pə'teiniŋ] a. 附属……的，与……有关的，为……所固有的
pesticide ['pestisaid] n. 杀虫剂，农药
petrochemical [,petrəu'kemikl] a. 石油化学的
petrolatum [petrə'leitəm] 石蜡油，软石蜡，矿酯
pharmaceutical [fɑ:mə'sju:tikəl] a. 药的，制药的
phase [feis] n. 相，状态
phenol ['fi:nɔl] n. 酚醛树脂
phenolics [fi'nɔliks] n. 酚醛塑料（树脂）
phosphoric [fɔs'fɔrik] a. 磷的，含(五价)磷的
phthalate [f'θæleit] n. 邻苯二甲酸盐（酯）
phthalic anhydride 邻苯二甲酸酐
physical vapor deposition 物理气相淀积
pictogram ['piktə,græm] n. 皮克，微微克
pigments ['pigmənt] n. 色素，颜料
planar ['pleinə] a. 平面的，平坦的
platinum ['plætinəm] n. 铂，白金（Pt）
plump [plʌmp] vi. 投票赞成，坚决拥护（for）
plywood ['plaiwud] n. 胶合板
pneumatic [nju:'mætik] a. 气力的，气动的
pneumatic dryer 气流干燥器
polar ['pəulə] a. 极性的
pollutant [pə'lu:tənt] n. 污染物
pollution [pə'lu:ʃən] n. 污染
polyurethane [,pɔli'juərəθein] n. 聚氨基甲酸酯
polyvinyl [pɔlivainl] n. 乙烯基聚合物
poly vinyl chloride 聚氯乙烯
polyampholyte [,pɔli'æmfə,lait] n. 聚两性电解质
polybutylene terephthalate 聚丁烯对苯二酸酯
polycarbonate n. 聚碳酸酯
polyelectrolyte [,pɔlii'lektrəu,lait] n. 聚合电解质
polyoxymethylene n. 聚甲醇，聚氧化亚甲基
polyphenylene oxide 聚苯氧化物
polystyrene [pɔli'staiəri:n] 聚苯乙烯

polyurethane [,pɔli'juəriθein] n. 聚氨酯，聚氨基甲酸酯
porosity [pɔ:'rɔsiti] n. 多孔性，有孔性，空隙率
porous ['pɔ:rəs] a. 多孔的
pose [pəuz] v. 造成，形成，提出
potential [pə'tenʃəl] n. （势）能，位（能），电势（位，压）
potential flow 势流
practice ['præktis] n. 实践操作
precipitate [pri'sipiteit] n. 沉淀物；v. 沉淀
precursor [pri'kɜ:sə] 产物母体，前身，先驱
predominate [pri:dɔmineit] vi. 占优势，居支配地位
prefer [pri'fə:] n. 宁愿（选择），更喜欢
prefix ['pri:fiks] n. 前缀，（人名前的）称谓（如 Mr.Dr.Sir 等）
prefix ['pri:fiks] n. 前缀
preliminary [pri'liminəri] a. 初始的，初步的
primary ['praiməri] a. 主要的，基本的（多相的）
prime [praim] a. 最初的；基本的
product ['prɔdʌkt] n. 产物
promote [prəu'məut] v. 增进，促进
propagate ['prɔpəgeit] v. 传播，蔓延
propagation [prɔpə'geiʃən] n. 增长，繁殖，传播，波及
propane ['prəupein] n. 丙烷
propel [prə'pel] vt. 推进，驱使
protein ['prəuti:n] n. 蛋白质
Protland cement 硅酸盐水泥，波兰特水泥，普通水泥
proton ['prəutɔn] n. 质子
protonate v. 使质子化
prototype ['prəutətaip] n. 原型，主型
proven ['pru:vn] a. 验证的
p-tolualdehyde n. 对甲苯甲醛
pucker ['pʌkə] v. 折叠；n. 皱纹
pulp [pʌlp] n. 浆状物，纸浆；矿浆
purity ['pjuəriti] n. 纯度，纯洁
pursuit [pə'sju:t] n. 追求，从事，研究
pyrolysis [pai'rɔlisis] n. 热解（作用），高温分解
quantum ['kwɔntəm] n. 量子
quantum theory 量子论
quartz [kwɔ:z] n. 石英，水晶
quicklime ['kwiklaim] n. 生石灰，氧化钙
quicklime ['qwiklaim] n. 生石灰，氧化钙
radiator ['reidieitə] n. 辐射体，散热器，暖气装置
radical ['rædikəl] n. 基，原子团；根部；根式
raffinate ['ræfineit] n. 提余液，残液
randam ['rændəm] a. 随机的，无规则的，偶然的

rate law 速度定律
rationalization ["ræʃənəlaiˈzeiʃən] n. 合理化
rayon [ˈreiɔn] n. 人造丝，人造纤维
reaction injection moulding 反应注射成型
reactive processing of molten polymers 熔融聚合物的反应加工
reboiler [riˈbɔilə] n. 再沸器
recorder [riˈkɔ:də] n. 记录仪，记录员，录音机
recorder [riˈkə:de] n. 记录器
recover [riˈkʌvə] vt. 回收，复原
rectify [ˈrektifai] vt. 精馏，精炼，蒸馏
rectifying section 精馏段
refer to … as … 把……称作
reflux [ˈri:flʌks] ad. 回流，倒流
refractory [riˈfræktəri] a. 难熔的，耐火的；n. 耐火材料
regulator [ˈregjuleitə] n. 调节器，稳定器
relative volatility 相对挥发度（性）
relay [ˈri:lei] vt 传递
remediation n. 补救,修补;治疗
reminiscent [remiˈnis(ə)nt] a. 回忆往事
replication [repliˈkeiʃən] n. 重复实验
representative [repriˈzentətiv] n. 代表，典型
reproducibility n. 再生性，还原性，重复性
reproducible [,ri:prəˈdju:səbl] a. 可再生的，可复写的
repulsion [riˈpʌlʃən] n. 推斥，排斥，严拒
requisite [ˈrekwizit] n. 必要的
residence time 停留时间
residue [ˈreizidju:] n. 残余物，残渣，剩余物
resilient [riˈziliənt] a. 有弹性的,能恢复原状的
respiratory [rəˈspirətri] adj. 呼吸的
restrict to 把……局限于……（范围）
restrict [risˈtrikt] v. 局限，限制性
retard [riˈtɑ:d] vt. 延迟，使减速，阻止，阻碍
retention [riˈtenʃən] n. 保留 保持
retention time [riˈtenʃən taim] 保留时间
revenge effect 报复效应
reverse reaction 逆反应
Reynolds number 雷诺数
rhodium [ˈrəudiəm] n. 铑（Rh）
road oil 铺路沥青
robust [rəuˈbʌst] a. 加强的，增强的，健全的
rotary drum dryer 转鼓式干燥器
rotary dryer 旋转干燥器
sacrificial [sækriˈfiʃəl] a. 牺牲的
sake [seik] n. 缘故
salable [ˈseiləbl] a. 畅销地，销路好的

sample [ˈsɑ:mpl] n. 试样
saturated [ˈsætʃəreitid] a. 饱和的
scatter [ˈskætə] v. 散射，分散，散开
scavenger [ˈskævindʒə] n. 清除剂，净化剂
schematic diagram [skiˈmætikˈdaiəgræm] 示意图
scout [skaut] vt.；vi. 侦察，搜索，监视，发现，排斥
scraped-surface crystallizer 刮膜式结晶器
screen [skri:n] n. 筛，网
screening [ˈskri:niŋ] n. 筛选，甄别
scrub [skrʌb] v. 使（气体）净化，洗气，洗涤
scrubber [ˈskrʌbə] n. 洗涤器
seal [si:l] n. 封铅，封条 图章 密封；vt. 封，密封
seam [si:m] n. 煤层
secondary [ˈsekəndəri] 仲[指 CH_3—$CH(CH_3)$-型支链烃，或指二元胺及 R_2CHOH 型的醇]
secondary reformer 二段（次）转化炉（器）
sediment [ˈsedimənt] n. 沉积物;沉积,沉淀
segment [ˈsegmənt] n. 部分；切片
segregation [segriˈgeiʃən] n. 分离，分凝，分开
self-heated steam gasification 自热式水蒸气气化
semipermeable [,semiˈpə:miəbl] a. 半透性的
sense [sens] vt. (自动)检测
sensitivity [sensiˈtiviti] n. 敏感度，灵敏度
separate [ˈsepəreit] v. 分离，分开
shaft [ʃɑ:ft] n. （传动，旋转）轴
shear stress 剪应力
shed [ʃed] vt. 放射，散发，流出，流下
shed light on 阐明，把……弄明白
shell [ʃel] n. 壳
shell and tube heat exchanger 管壳式换热器，列管式换热器
shift reaction 变换反应，转移反应
shikimate [ˈʃikmeit] n. 莽草酸盐
short tube evaporator 短管蒸发器
sidestream n. 侧线馏分，塔侧抽出物
sieving [ˈsiviŋ] n. 筛选，筛分法
sieving [siviŋ] n. 筛分
significant [sigˈnifikənt] a. 有意义的，意味深长的，重要的
silica [ˈsilikə] n. 二氧化硅
silica-gel （氧化）硅胶
silicate [ˈsilikit] n. 硅酸盐（酯）
silicone rubber [silikən ˈrʌbə] 硅橡胶
simplicity [simˈplisiti] n. 简单，简易
simulator [ˈsimjuleitə] n. 模拟器
simultaneously ad. 同时地；同时发生地
skeleton [ˈskelitn] n. 骨架，骨骼，基干，构架

slag [slæg]　n. （炉，熔，矿）渣
slake　v. 消解, 潮（水）解
slakedlime　n. 熟石灰, 消石灰
sludge [slʌdʒ]　n. 淤泥, 泥状沉积物; 淤渣
slurry polymerization　淤浆聚合
soda [ˈsəudə]　n. （钠）碱, 苏打, 碳酸钠
ash [æʃ]　n. 灰, 粉
sodaash　纯碱, 苏打灰, 碳酸钠
solar pan　盐池, 盐田
solubility [sɔljuˈbiliti]　n. 溶解度, 溶解性
solubility product　溶度积
solution polymerization　溶液聚合
solvent [ˈsɔlvənt]　n. 溶剂, a. 有溶解力的
solvent naphtha　溶剂石脑油
sophistication [səfistiˈkeiʃən]　n. 完善, 改进, 复杂化
sorbent [ˈsɔːbənt]　n. 吸附剂, 吸收剂
spaghetti [spəˈgeti]　n. 通心粉
sparging [ˈspɑːdʒiŋ]　n. 起泡, 鼓泡, 喷射,
specific value　比值
spontaneous [spɔnˈteinjəs]　a. 自发的, 自然的
spray [ˈsprei]　n. 喷雾, 喷淋; v. 喷
spray dryer　喷雾干燥器
spurt [spəːt]　vt.; vi. 喷射出, 急剧上升, 生长, 发芽
stab [stæb]　vt.; vi. 刺(穿, 伤), 企图, 努力, 尝试
standby [ˈstændbai]　a. 备用的, 后备的; n. 备用设备
standpoint [ˈstændpɔint]　n. 立场; 观点
　from the standpoint of ……站在……的立场上; 根据……的观点
stationary [ˈsteiʃ(ə)nəri]　a. 固定的, 不动的; n. 固定
statistical mechanics [stəˈtistikəl miˈkæniks]　统计力学
steam ejector　蒸汽喷射器
stem [stem]　n. 茎, 柄
stiffness [ˈstifnis]　n. 刚性度, 韧性
still [stil]　n. 蒸馏釜, 蒸馏
stirred tank reactor　搅拌釜式反应器
store [stɔː]　vt. 储藏, 储备; 装备
straight run gasoline　直馏汽油
stratosphere [ˈstrætəusfiə]　n. 同温层, 平流层
striking [sˈtraikiŋ]　a. 引人注目，显著的
strip [strip]　vt. 解吸, 汽提
stripper [ˈstripə]　n. 汽提塔, 解吸塔
stripping [ˈstripiŋ]　n. 洗提, 气提, 解吸
stripping section　提馏段

strive [straiv]　vi. 努力, 奋斗, 力求, 斗争, 反抗
styrofoam [ˈstairəfəum]　n. 泡沫聚苯乙烯
subject [ˈsʌbdʒikt]　vt. 遭受, 蒙受（to）
submerged combustion　浸没燃烧
subsequently [ˈsʌbsikwəntli]　ad. 其后, 其次
subsidence [ˈsʌbsidəns]　n. 沉淀, 陷没, 下沉
substituent [sʌbˈstitjuənt]　n. 取代, 取代基; a. 取代的
subterranean [ˌsʌbtəˈreiniən]　a. 地下的, 隐藏的, 秘密的
subunit [ˈsʌbˌjuːnit]　n. 副族, 子单元, 亚组, 子群
suck [sʌk]　v. 吸入
suffix [ˈsʌfiks]　n.; vt. 后缀, 下标, 添后缀
suffix [ˈsʌfiks]　n. 后缀
sulfate [ˈsʌlfeit]　n. 硫酸盐
sulfide [ˈsʌlfaid]　n. 硫化物
sulfur=sulphur
sulfuric [sʌlˈfjuərik]　a. 硫的, 硫黄的
sulphuric [sʌlˈfjuərik]　a. 硫的
sulphuric acid　硫酸
superimpose [sjuːpərimˈpəuz]　v. 附加, 加在上面
supersaturation [ˈsjuːpəˌsætəˈreiʃən]　n. 过饱和（现象）
supersede [sjuːpəˈsiːd]　vt. 代替, 取代, 废弃
support [səˈpɔːt]　n. 保障
surfactant [səˈfæktənt]　n. 表面活性剂
susceptible [səˈseptəbl]　a. 易受影响的; n. （因缺乏免疫力而）易得病的人
suspect [səsˈpekt]　vt. 猜想, 怀疑
suspension [səˈspenʃən]　n. 悬浮液
suspension polymerization　悬浮聚合
suspicion [səsˈpiʃən]　n. 猜想, 怀疑
symposia [simˈpəuziə]　n. 专题报告会, 讨论会, 专题文集
synonymously [siˈnɔniməsli]　ad. 同（意）义地
synthetic [sinˈθetik]　a. 合成的
systematic [ˌsistiˈmætik]　a. 体系的, 系统的
tank crystallizer　槽式结晶器
tantalum [ˈtæntələm]　n. 钽
tar [tɑː]　n. 焦油
Teflon [ˈteflɔn]　n. 特氟隆, 聚四氟乙烯
tension [ˈtenʃən]　n. 张力, 弹力
tentative [ˈtentətiv]　a. 不明确的
term [təːm]　n. (比例或方程的)项
test [test]　vt. 检验, 验证, 试验
tetra-alkyl lead　四烷基铅
tetraethyl lead　四乙基铅
tetrahedral [ˌtetrəˈhedrəl]　a. 四面体的

tetrahedron ['tetrə'hedrən] n. 四面体
tetravalent [,tetrə'veilənt] a. 四价的; n. 四价染色体
the Solvay process 索尔维法
thermal tubular reactor 热管反应器
thermally ad. 用热的方法
thermochemical ['θə:mou'kemikəl] a. 热化学的
thermodynamics ['θə:moudai'næmiks] n. 热力学
thixotropy [θik'sɔtrəpi] n. 触变性
threefold ad. 三倍（于），增量三倍
thumb one's noise (at) （对……）作蔑视的手势
tilt up 翘起
titled a. 倾斜的，翘起的
transition [træn'siʃən] n. 过渡（段），转变，变化
translation [træns'leiʃən] n. 转化，变换
transparent [træns'pɛərənt] a. 透明的，半透明的，某种辐射线可以透过的
transport [træns'pɔ:t] vt. 运输，输送，运送; n. 运输，输送，运送
tray dryer 盘架干燥器
trial ['traiəl] n. 试验，审判
trichloroethene 三氯乙烯
trickle bed reactor 滴流床反应器
triene ['traii:n] n. 三烯
trinitrotoluene [trai'naitrəu'tɔljui:n] n. 三硝基甲苯，TNT 炸药
trivial ['triviəl] a. 普通的，不重要的，无价值的
troposphere ['trɔpəsfiə] n. 对流层
tube sheet 管板
tubular ['tju:bjulə] a. 管的,管式的,由管构成的
tubular fixed-bed reaction 管式固定床反应器
tubular reactor 管式反应器
turbulent ['tə:bjulənt] a. 湍流的，紊流的; 扰动的
U.S. PTO=United States Patent Office 美国专利局
ultimately ['ʌltimitli] ad. 最后，最终
ultraviolet [,ʌltrə'vaiəlet] a. 紫外的,紫外线的
unambiguous [ʌnæm'bigjuəs] a. 明确的，清楚的
unbound electron 自由电子
underlain [,ʌndə'lein] vt. 作为……的基础
underlay [,ʌndə'lei] vt. 作为……的基础
underlie [,ʌndə'lai] vt. 作为……的基础
underlying [,ʌndə'laiiŋ] a. 作为……的基础
unidirectional ['ju:nidi'rekʃənl] a. 单向性的
uniformity [,ju:ni'fɔ:miti] n. 一致性；均匀性；均一性
unimpressive [,ʌnim'presiv] a. 给你印象不深的，平淡的，不令人信服的
unintended consequence 缺乏研究（知识，经验，外行）的结果(结论)
unlagged a. 未保温的，未隔热的，未绝缘的
unpaired [,ʌn'pɛəd] a. 不成双的，无对手的
untapped a. 未利用的，未开发的
uranium [juə'reiniəm] n. 铀
use up 消耗掉，用完
valence ['veiləns] n. （化合）价，原子价
validity [və'liditi] n. 有效，合法性；正确，确实
valve [vælv] n. 阀
vanadium [və'neidjəm] n. 钒（V）
vaporize ['veipəraiz] v. 汽化
vapor-phase chromatography ['veipə 'feiz krəumə'tɔgrəfi] 气相色谱（法）
vehicle ['vi:ikl] n. 车辆
verifiable ['verifaiəbl] n. 可核实的
versatile [,və:sətail] ① 多方面的，多才多艺的; ② 万用的，通用的
versatility [,və:sə'tiliti] n. 多功能性
vertical ['və:tikl] a. 垂直的
virgin oil 直馏油
virtue ['və:tʃu:] n. 效能，效力，美德
viscous ['viskəs] a. 黏（性，滞，稠）的
visionary ['viʒə,neri:] a.; n. 幻想的，空想，非实际
visualization [,viʒuəlai'zeiʃən] n. 是看得见的
vital ['vaitl] a. 必需的，生命的
viz: (videlicet [vi'di:liset] 之略) 即，就是
void [vɔid] n. 空隙，空隙率；空间，空位
volatility [vɔlə'tiliti] n. 挥发度，挥发性
vulcanization ['vʌlkənai'zeiʃən] n. 硫化，硬化
wash [wɔʃ] v. 洗涤，洗
water gas 水煤气
wet [wet] vt. 把……弄湿; a. 湿
wiped-film evaporator 刮膜式蒸发器
wise [waiz] a. 明智的，考虑周到的，慎重的
xerogel ['ziərədʒel] n. 干凝胶
zinc [ziŋk] n. 锌（Zn）
zirconium [zə'kəunjəm] n. 锆
zwitterionic [,zwitərai'ɔnik] a. 两性离子的